EASY PIC'n

A BEGINNER'S GUIDE TO USING
PIC16/17 MICROCONTROLLERS
FROM

DAVID BENSON

VERSION 2.0

NOTICE

The material presented in this book is for the education and amusement of students, hobbyists, technicians and engineers. Every effort has been made to assure the accuracy of this information and its suitability for this purpose. Square 1 Electronics and the author assume no responsibility for the suitability of this information for any application nor do we assume any liability resulting from use of this information. No patent liability is assumed for use of the information contained herein.

TRADEMARKS

PIC is a registered trademark of Microchip Technology Inc. in the U.S.A.
PICSTART is a trademark of Microchip Technology, Inc.

MS-DOS is a registered trademark of Microsoft Corporation.

ISBN 0-9654162-0-8

PUBLISHER

Square 1 Electronics
P.O. Box 501
Kelseyville, CA U.S.A.

Voice (707)279-8881
FAX (707)279-8883
EMAIL squareone@zapcom.net

A BEGINNER'S GUIDE TO USING PIC16/17 MICROCONTROLLERS FROM

INTRODUCTION

The PIC16/17 was originally designed as a Peripheral Interface Controller (PIC) for a 16-bit microprocessor. It was essentially an I/O controller and was designed to be very fast. It had a small microcoded instruction set. This design became the basis for Microchip Technology's PIC16C5X family of microcontrollers.

The PIC16C84 microcontroller is unique because its program memory is EEPROM. It can be programmed, tested in-circuit and reprogrammed if necessary in a matter of a few minutes and without the need for a UV EPROM eraser. It is a small device (18-pin), readily available to all including hobbyists and students at a cost of $6.50 (at this writing) in single quantity.

Think of the PIC16C84 as a custom I/O handler. It looks at its inputs and, based on what it sees, it sends signals out its outputs. You can customize it to do what you want via programming. It is not a heavy duty number cruncher or data manipulator.

The MPASM assembler from Microchip Technology (manufacturer of PIC16/17 family) is the PIC16/17 assembler of choice for use in this book because:

- It is free from Microchip
- Most people who are into PIC16/17's speak "MPASM"
- Most examples in magazines and on the Internet are written in the MPASM dialect

A variety of programmers are available for the PIC16C84 (see sources in Appendix A). Some range in cost from $40 to $70. Some come with their own assemblers but will accept hex files created by MPASM (details later). Microchip makes several PIC16/17 programmers at this writing. The PICSTART 16B1 is the low end model which will program base-line PIC16C5X (18 and 28-pin) and 18-pin mid-range PIC16C61, PIC16C62X, PIC16C71, PIC16C84 microcontrollers. The PICSTART 16B1 is used as the example in this book.

The PIC16C84, MPASM assembler and a low cost PIC16C84 programmer make a very inexpensive development system and a great way to get started creating microcontroller based projects.

What you learn using the PIC16C84 is directly applicable to the whole line of PIC16/17 microcontrollers. The approach taken with the PIC16C84 in the first part of book is backwards compatible with the PIC16C54 (explained toward the end of the book). The hassles of file register bank switching and program memory paging are avoided as they are confusing to a beginner but manageable after some experience is gained.

Learning how PIC16/17's work and how to apply them involves study in three areas.

- MS DOS computer use (as needed)
- Assembler
- PIC16/17 itself

PIC16/17's are not easy to write code for by hand because they are not intuitive. Many instructions require setting or clearing a bit to specify the destination for the result of executing the instruction. This is ok as an initial learning experience, but would become tedious very soon. So use of "power tools" is essential. This means learning to use an MS-DOS computer if you haven't already done so. It also means learning to use an assembler which converts English-like readable instructions into machine language understood by the PIC16/17 itself.

Finally, learning about the PIC16/17's inner workings is possible once use of the power tools is understood.

The object is to make this process as easy and enjoyable as possible. Once you get through this and you have programmed a PIC16C84 for the first time, a whole new world awaits. You will be able to create more interesting projects and have more fun!

The usual approach to teaching the use of the PIC16/17 is to assume total knowledge of MS-DOS, get into all of the assembler commands and then show advanced examples. There is no simple example showing how to get started. As usual, only 5 percent of this information is needed to get started, but which 5 percent. The approach taken here will be to give you the 5 percent you need to get going.

The assumption is made that you know how to do the following on an MS-DOS computer:

- Create a new directory called \PIC
- Copy the contents of a floppy disk to the directory \PIC
- Use a simple text editor to create a text file, save it, make a copy of it, print it, and copy it to the directory \PIC

PIC16/17 PRODUCT OVERVIEW

This book uses the PIC16C84 and later the PIC16C54 as examples to get you started. The emphasis is on the base-line (12-bit core) and mid-range (14-bit core) products. The 12-bit core designates the instruction word length. You really won't care how many bits there are in an instruction word, but it is the basis on which Microchip divides up their product line. The following brief tables are for reference:

	Base-Line	Mid-Range
	12-bit Core	14-bit Core
Stack Levels	2	8
Interrupts	No	Yes
Instructions	33	35*
Serial Programming	No	Yes
File Registers	128 max **	256 max **
Example	PIC16C54	PIC16C84

* We will use 37 - details later
** Varies with part number

Part No.	Feature
PIC16C6X	Digital I/o
7	Analog Input
8	EEPROM
9	Combination

At this writing, Microchip has announced that two new 8-pin microcontrollers will be available soon which use the PIC16C5X core. This will make even more low-end microcontroller projects and products possible.

Also, Microchip is in the process of replacing the PIC16C84 with a new version, the PIC16F84. For our purposes, they are essentially identical.

PIC16C84

PINS AND FUNCTIONS

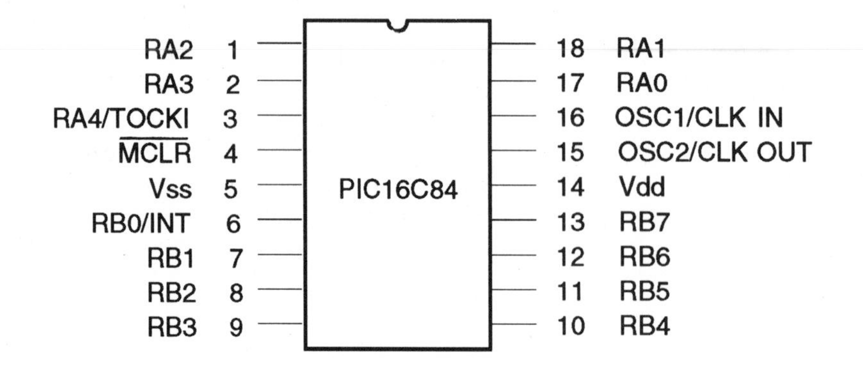

The PIC16C84 is fabricated using CMOS technology. It consumes very little power and is fully static meaning that the clock can be stopped and all register contents will be retained. The maximum sink and source current by any I/O pin is 25 mA and 20 mA respectively. The maximum sink or source current for a port at any given time is as follows:

	Port A	Port B
Sink current	80 mA	150 mA
Source current	50 mA	100 mA

Supply current is primarily a function of operating voltage, frequency and I/O pin loading and is typically 2 mA or so for a 4MHz clock oscillator. This drops to approximately 40 microamps in the sleep mode.

Because the device is CMOS, all inputs must go somewhere. All unused inputs should be pulled up the supply voltage (usually +5VDC).

PACKAGE

The PIC16C84 is available in an 18-pin DIP package suitable for the experimenter. The part number is PIC16C84-04/P for the 4 MHz part.

CLOCK OSCILLATOR

Four different types of clock oscillators may be used with PIC16C84 parts in general. The clock oscillator types are:

RC - resistor/capacitor
XT - crystal or ceramic resonator
HS - high speed crystal or ceramic resonator
LP - low power crystal

At programming time, the part must be told via so-called configuration "fuses" what type of oscillator it will be connected to.

The details of various oscillator circuits and components are given in the Microchip data book.

My suggested clock circuit for experimentation is a 4 MHz crystal-controlled oscillator in a can. They are accurate, always start and the math required for designing time delay loops is simple. The external clock frequency is divided by 4 internally, so the internal clock frequency is 1 MHz. Almost all instructions are executed in one instruction (internal clock) cycle or 1 microsecond in this case. This is plenty fast for experimenting. The oscillator type code is XT.

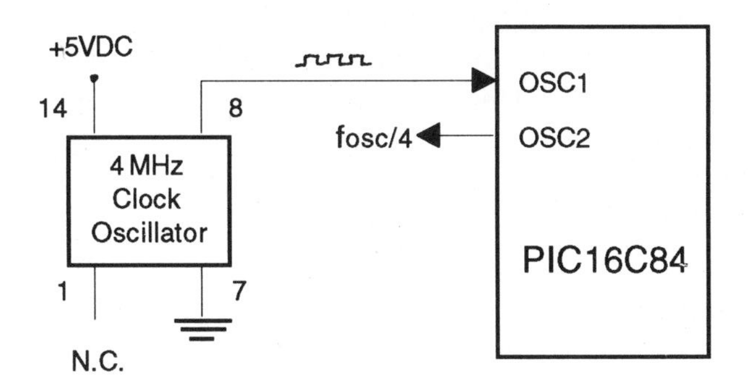

Ceramic resonators with built-in capacitors are small and inexpensive while providing good accuracy (+ 1.3 percent or better). They look a lot like a small ceramic capacitor except for the fact they have three leads. A typical device looks like this:

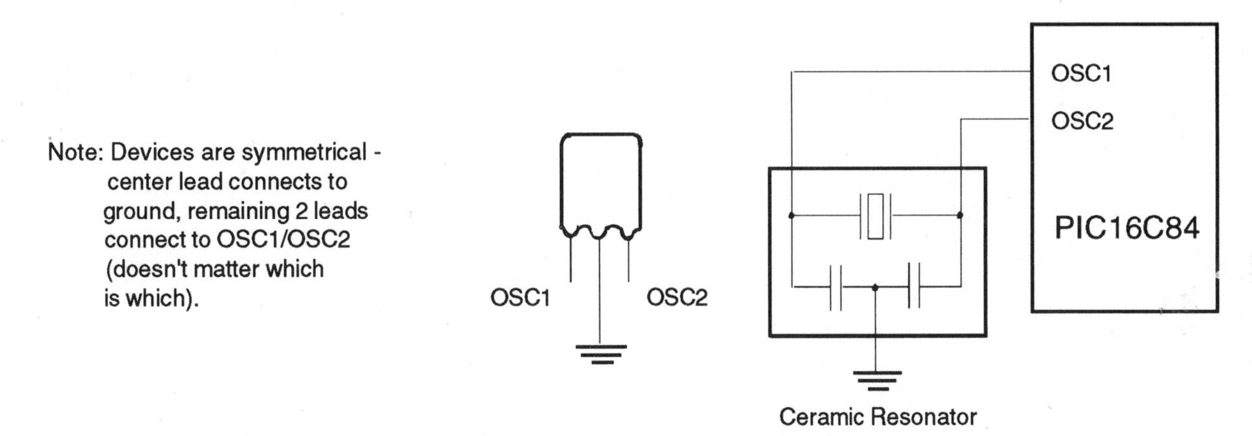

A standard frequency is 4.00 MHz which offers the advantages mentioned above. Use XT for oscillator type. Keep the wires/traces short.

An RC clock circuit may also be used where cost is an issue and timing accuracy is not. Examples are shown. The oscillator type code is RC.

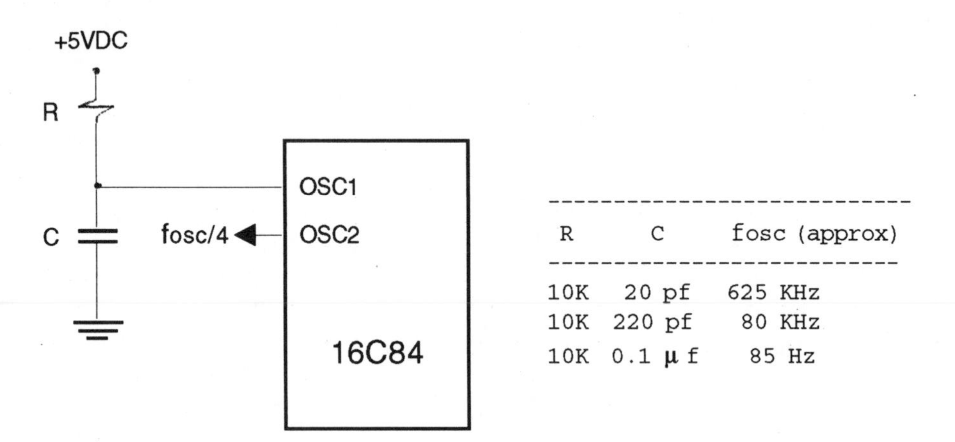

R	C	fosc (approx)
10K	20 pf	625 KHz
10K	220 pf	80 KHz
10K	0.1 µf	85 Hz

CONFIGURATION "FUSES"

The configuration EEPROM is not part of the EEPROM used for program storage. The data book and assembler screen refer to fuses, but there aren't any fuses as such. There are four EEPROM bits. Two select clock oscillator type, one is the watchdog timer enable bit, and one is the code protection "fuse".

RESET

The PIC16C84 has built-in power-on reset which works as long as the power supply voltage comes up quickly. Commonly the $\overline{MCLR}$ pin is merely tied to the power supply. A switch may be used to regain control if things run away.

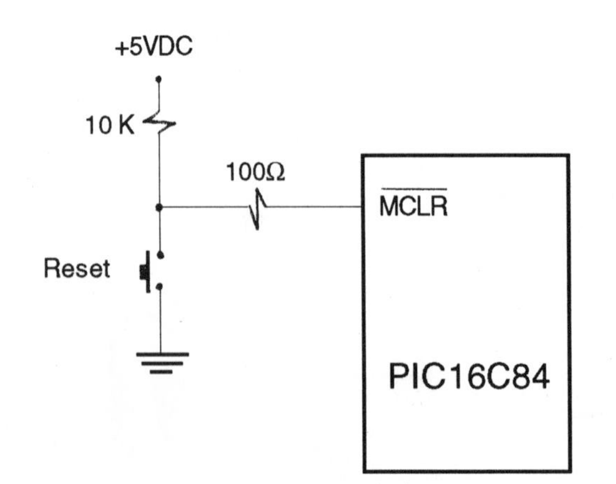

PORTS

Port A is 5 bits/lines wide and port B is 8 bits/lines wide or byte-wide. Each port line may be individually programmed as an input line or output line. This is done using a special instruction which matches a bit pattern with the port lines. A "0" associated with a port line makes it an output, a "1" makes it an input. Examples follow.

The PIC16C84 has port data direction registers as many other microcontrollers do. The PIC16C54 does not (more on this later). Using the port data direction registers involves file register bank switching which I have chosen to avoid for now. Doing it in the manner I am about to describe goes against the recommendation of Microchip (concerned about compatibility with future products). My primary concern is to make learning about PIC16/17's as simple as possible. So the initial programming examples will use an obsolete instruction (TRIS) to avoid having to deal with bank switching at this early stage. A benefit of this is backward compatibility. You will need to learn to use this technique if you choose to work with base-line parts (PIC16C5X series).

The Port B lines have weak pullup resistors on them which may be enabled or disabled under software control. All 8 resistors are enabled/disabled as a group via the RBPU bit in the option register. The pullup resistor on an individual port line is automatically turned off when that line is configured as an output. The pullups are disabled on power-on reset.

Port A, bit 4 is shared with the external timer/counter input called T0CKI. As a digital input line, the input is Schmitt trigger. As a digital output line, it is open collector, so a pullup resistor is required. As an output, the logic is reversed. A "0" written to the port line outputs a logic 1. The output cannot source current, it can only sink current.

All unused port lines should be tied to the power supply (CMOS rule - all inputs must go somewhere). On reset, all port lines are inputs.

SPECIAL FEATURES

Watchdog Timer

The watchdog timer is useful in some control applications where a runaway program could cause a safety problem. We will not deal with it except to say that it is important to select "watchdog timer off" when programming the configuration "fuses".

Power-up Timer

The power-up timer should be selected "ON" when programming the configuration "fuses".

Sleep Mode

The feature of the "sleep mode" is drastically reduced power consumption achieved by turning off the main clock oscillator.

CIRCUIT FOR EXPERIMENTING WITH THE PIC16C54 AND PIC16C84

A circuit for experimenting with the PIC16C54 and PIC16C84 is shown. It has all the components required for most of the experiments in this book. Building it saves building one circuit for each experiment.

Providing a ZIF socket for the microcontroller is a good idea because it will be inserted and removed many times.

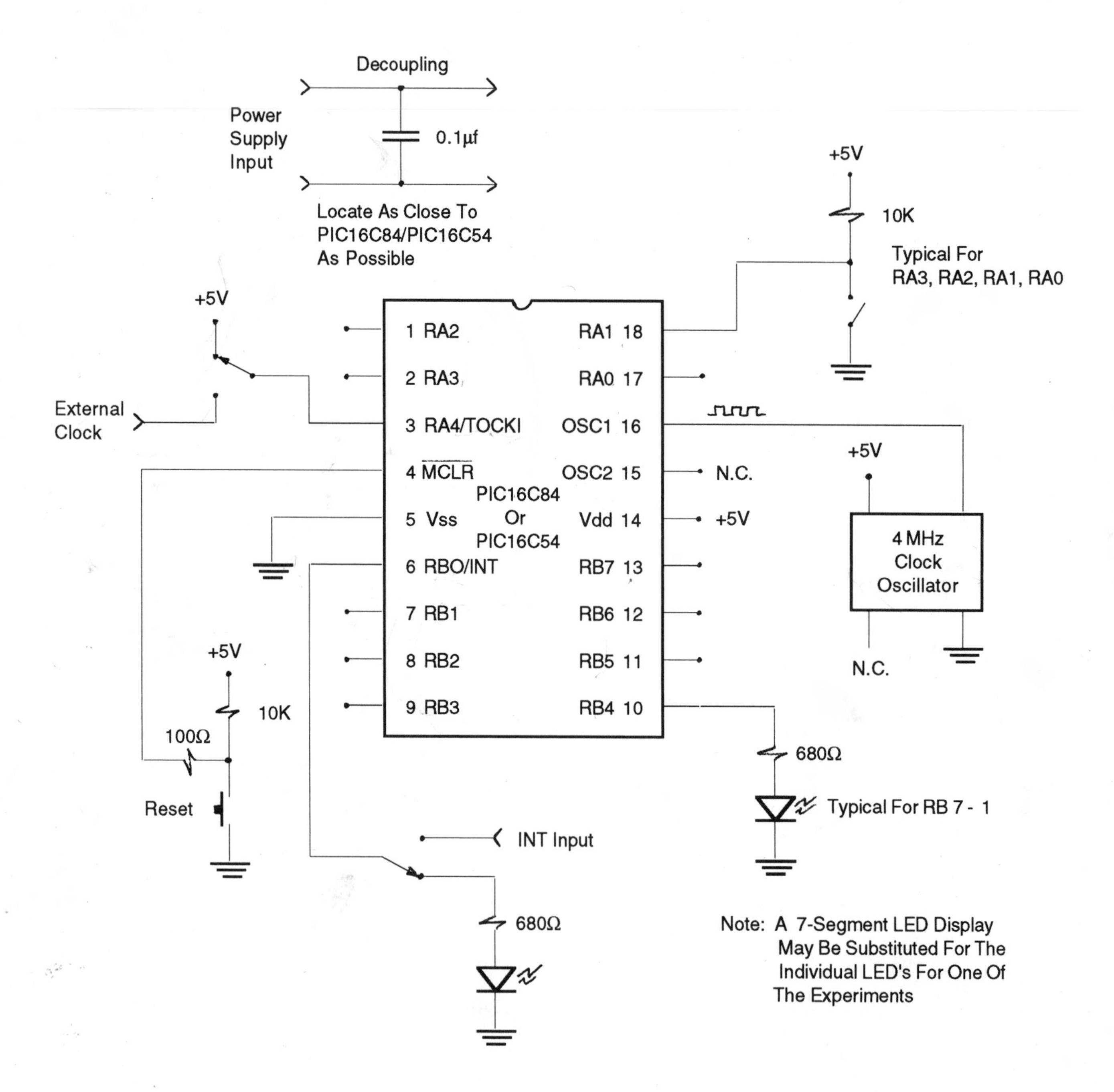

PIC16/17 ARCHITECTURE

PIC16/17 microcontrollers have two separate blocks of memory, program memory and file registers.

Program Memory

The PIC16C84 program memory is 14 bits wide and 1K words long. Program memory is EEPROM. Program memory is read-only at run time. PIC16/17's can only execute code contained in program memory.

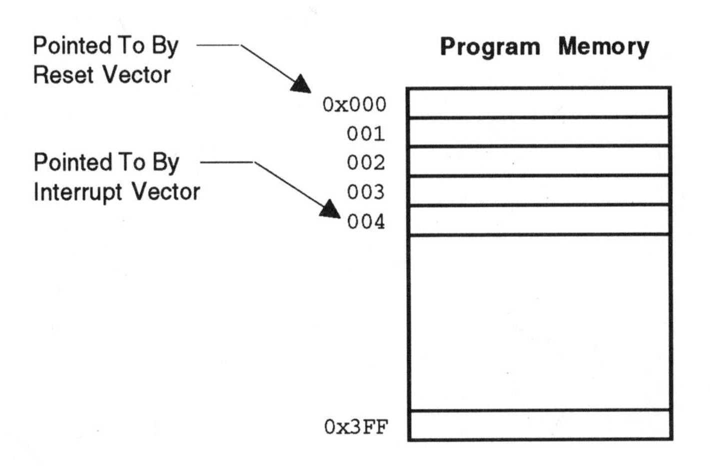

A limited amount of data may be stored in program memory as part of RETLW instructions (see programming chapter).

Weird Hex Notation

Some assembler listings show two-digit hexadecimal addresses in the form 0xXX where the upper case x's mean don't care. The "0x" means hexadecimal. The 0x notation comes from the C programming language. The main thing is, when you see 0x0F, it means hexadecimal 0F. 0x004 means hexadecimal 004.

File Registers

The file registers are 8 bits wide with the exception of the program counter which is 13 bits wide. The PIC16C84 has a 64 file register address space (0x00 - 0x2F), but not all addresses are used.

File Registers

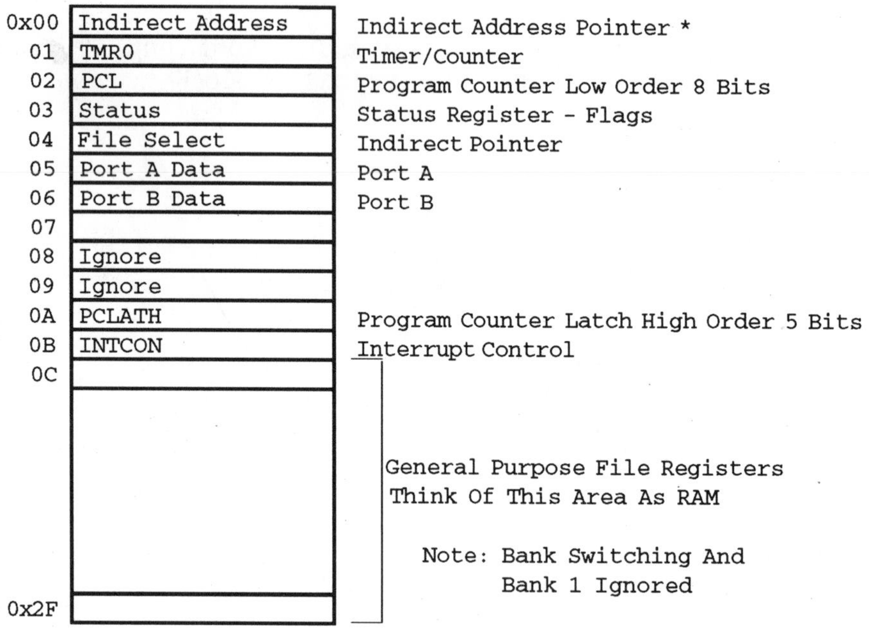

Address	Register	Description
0x00	Indirect Address	Indirect Address Pointer *
01	TMR0	Timer/Counter
02	PCL	Program Counter Low Order 8 Bits
03	Status	Status Register - Flags
04	File Select	Indirect Pointer
05	Port A Data	Port A
06	Port B Data	Port B
07		
08	Ignore	
09	Ignore	
0A	PCLATH	Program Counter Latch High Order 5 Bits
0B	INTCON	Interrupt Control
0C - 0x2F		General Purpose File Registers Think Of This Area As RAM Note: Bank Switching And Bank 1 Ignored

* Not Physically Implemented

Twelve file registers have specific dedicated purposes which will be of interest to you (to be described later). 36 file registers are there for your use and may be thought of as RAM or data memory for storing data during program execution.

	Hex Address	
f0	0x00	Indirect data addressing register. See indirect addressing section in programming chapter.
f1	0x01	Timer/counter register (TMR0). See timing and counting chapter.
f2	0x02	Program counter low byte. See relative addressing section in programming chapter.
f3	0x03	Status word register.
f4	0x04	File select register (FSR). See indirect addressing section in programming chapter.
f5	0x05	Port A - 5-bit, bits 5 - 7 are not implemented and read as 0's.
f6	0x06	Port B - 8-bit (bit 0 is also INT bit).
f7	0x07	Not used.
f8	0x08	Ignore - used in EEPROM programming.

f9	0x09	Ignore - used in EEPROM programming.
fA	0x0A	PCLATH - write buffer for upper 5 bits of program counter. See relative addressing section in programming chapter.
fB	0x0B	INTCON - interrupt control register. See interrupts chapter.
fC - f2F	0x0C - 0x2F	General purpose registers (RAM).

Note: File registers at 0x81, 0x85 and 0x86 are ignored for now. Bank switching is required to access them.

Working Register (W)

The PIC16C84 has an 8-bit working register called W. It is a lot like the accumulator in other microcontrollers. Its use will be illustrated in the programming chapter.

Option Register

The option register controls the port B weak pullup resistors, interrupt signal edge select, the prescaler for the timer/counter and watchdog timer (shared) and also the input to the timer/counter.

Stack

The PIC16C84 has an 8-level stack separate from the other registers (not part of the file register address space = data memory). It is used to save the program counter contents when subroutines are called so the microcontroller will know where to resume program execution when returning from the subroutine. This is also true of interrupts. There are no stack manipulation instructions and the stack is not accessible to the programmer.

Reset Vector

On reset which occurs on power-up or when the reset switch is used to pull $\overline{\text{MCLR}}$ low, the PIC16C84 will go to program memory location 0x000 where the first instruction is stored. It then begins executing instructions stored sequentially in memory.

Interrupt Vector

The interrupt vector points to 0x004, so if interrupts are used, the first instruction of the interrupt service routine will be at that location.

OPTION REGISTER

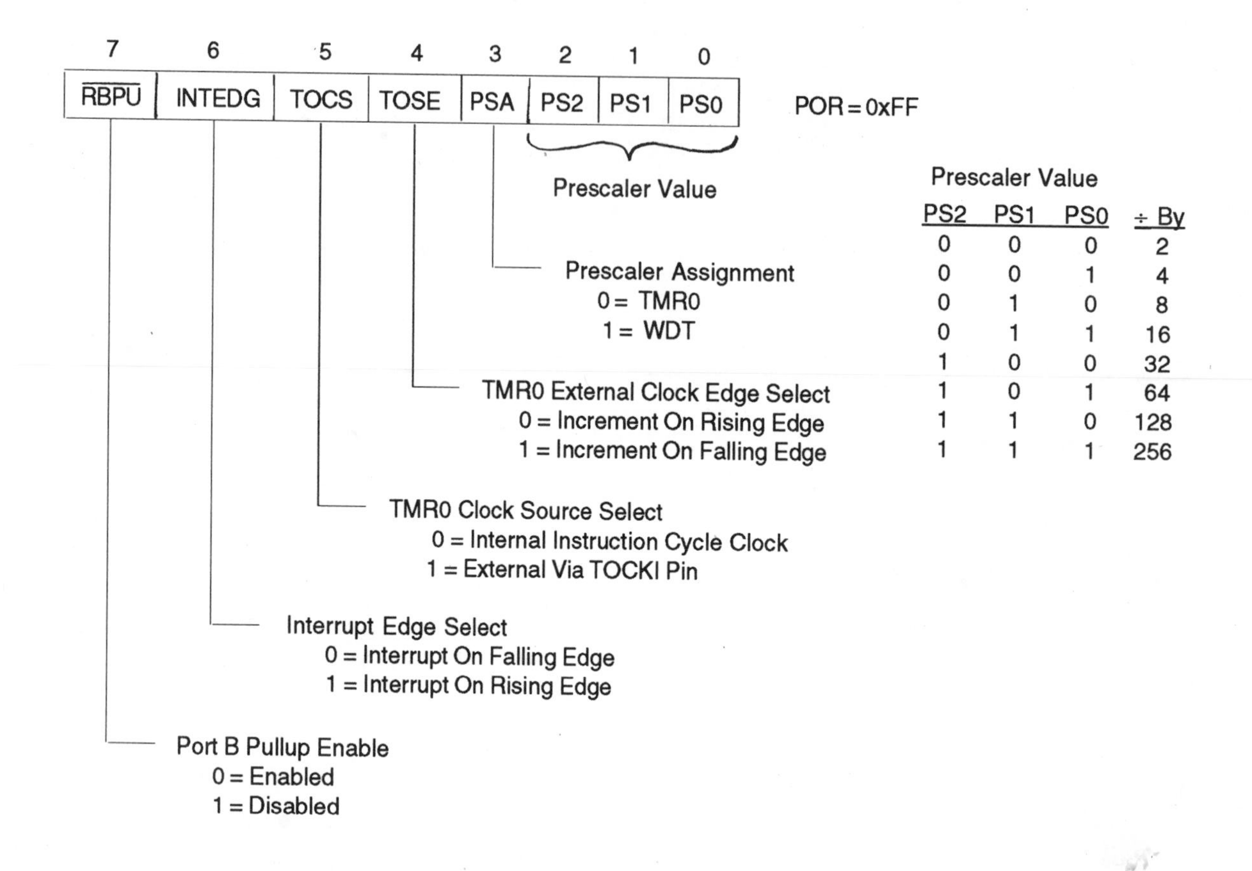

Prescaler Value

PS2	PS1	PS0	÷ By
0	0	0	2
0	0	1	4
0	1	0	8
0	1	1	16
1	0	0	32
1	0	1	64
1	1	0	128
1	1	1	256

For ease of use while learning and for backward compatibility with the PIC16C54, we will think of the PIC16C84 option register as a "buried" register (not included in the file register address space and having no address) and will write to it using the OPTION instruction. Microchip considers the option register obsolete, but we won't let that bother us. All of this avoids file register bank-switching which can be dealt with later.

To change the contents of the option register, we will use the OPTION instruction which loads the contents of the W register into the option register.

```
            movlw       b'00000000'   ;bit pattern defined
            option                    ;sends bit pattern to
;                                        option register
;
;-----------------------------------------------------
;           bit 7 = 0      port B pullups disabled
;           bit 6 = 0      interrupt on rising edge
;           bits 5,4,3,2,1,0 = 0 - details in
;                              timing and counting chapter
;-----------------------------------------------------
```

PROGRAM COUNTER

The PIC16C84 has a 13-bit program counter. GOTO and CALL instructions include an 11-bit address which is enough for a 2K program memory address space. The PIC16C84 has 1K of program memory.

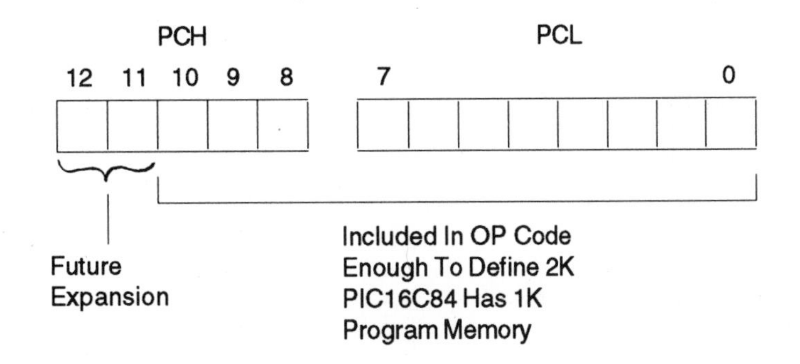

The program counter high (PCH) cannot be read or written to directly. It is loaded from a 5-bit latch called program counter latch high or PCLATH.

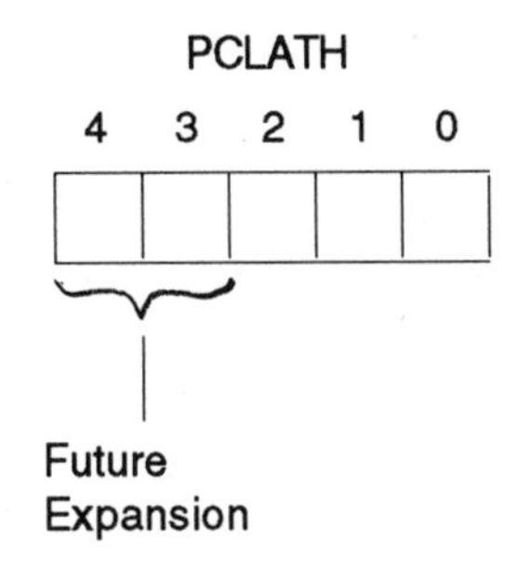

For the 14-bit core parts (PIC16CXX but not including PIC16C5X), a program memory page is defined as 2K (not 256 bytes as with many other 8-bit microcontrollers) because 2K can be accessed using the 11 address bits contained in GOTO and CALL instructions.

Since the PIC16C84 has 1K or one-half page of program memory, we won't need to contend with program memory paging. Avoiding this is a big help in getting started using PIC16/17's. To make this work, tables (which use a computed address) should not cross a 256 byte program memory block boundary because the PCLATH bits(s) are involved

With a knowledge of paging, a table can straddle a 256-byte block boundary. PCLATH is pre-conditioned and the destination address can be anywhere, but that stuff comes later.

For computed addresses which involve PCH, the bits come from PCLATH.

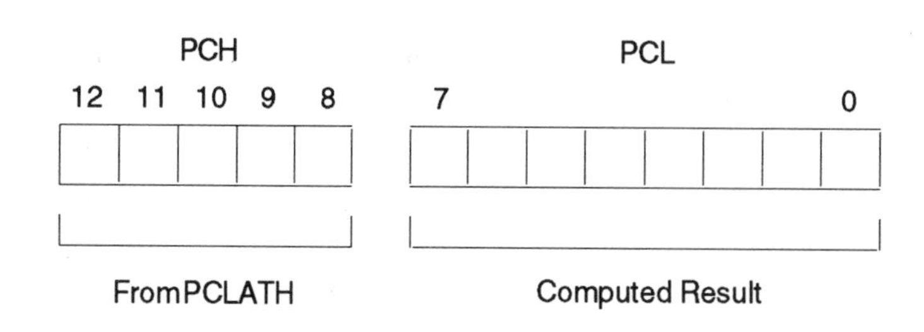

PCLATH comes up ---00000 on power-on reset, so you shouldn't have to be concerned after that unless your program writes to PCLATH. Bits 4,3 must be 0 and 0 (bits are for future expansion but must be 00 to make the PIC16C84 work properly).

Don't sweat these details for now. Their importance will come into focus after you have done some programming examples.

STATUS REGISTER

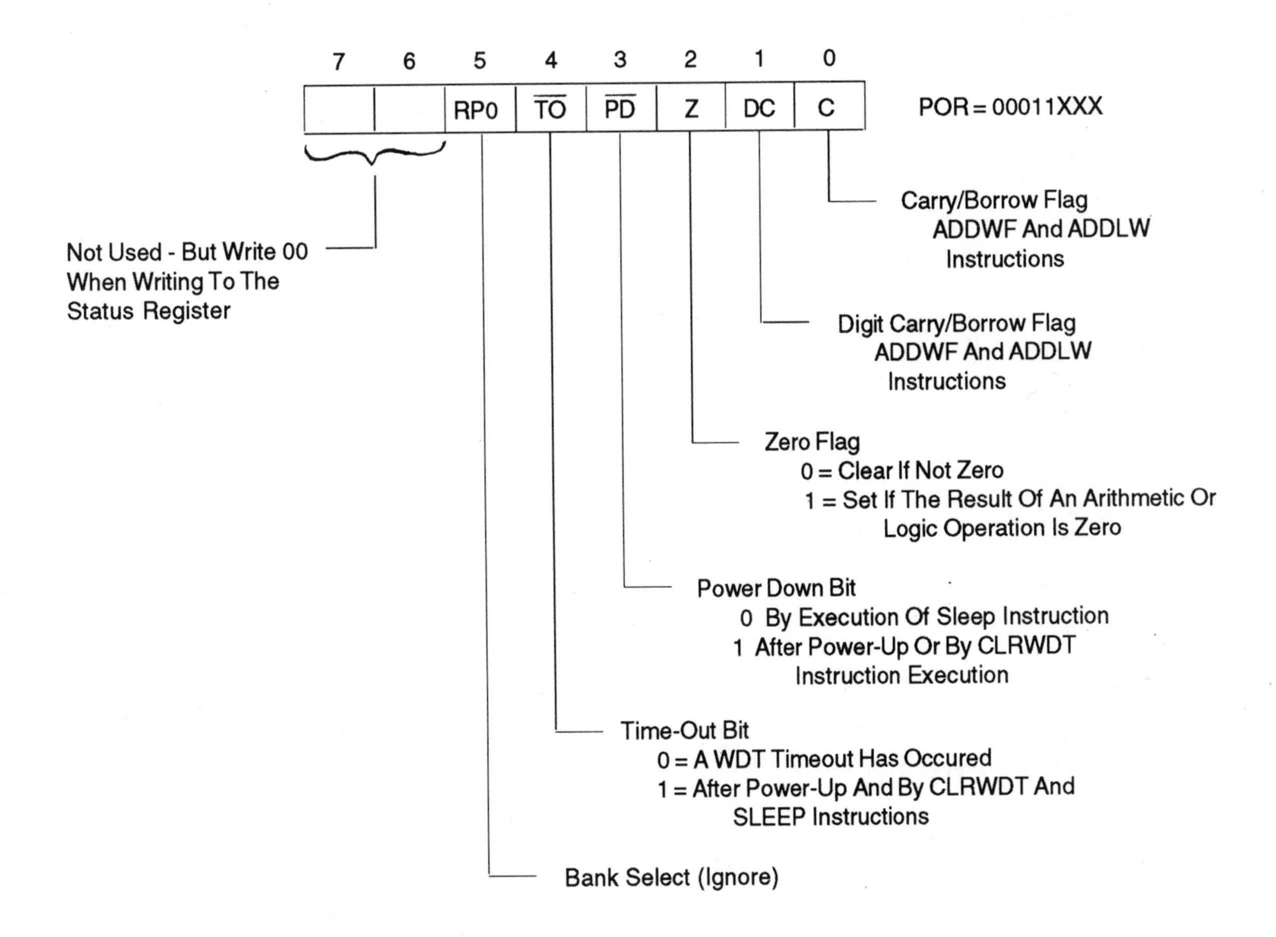

Care must be taken when reading the status register so as not to affect its contents. Using bit instructions (BTFSS or BTFSC) will not affect the contents while using MOVF does effect the Z flag.

Writing to the status register using BSF, BCF or MOVWF instructions will not effect the status register in any way other than the write itself (ie. no flags are affected by the execution of the instruction).

ASSEMBLERS

An assembler is a program which, in this case, runs on an MS-DOS computer and which converts readable instructions contained in a text file to hexadecimal code understood by the PIC16/17. The MPASM assembler from Microchip is fancy, has lots of capabilities, and it may be overkill for getting started. The advantages of using it are:

- It's free (obtainable from Microchip).
- Most people use it, i.e. most people speak MPASM.
- If you start out with it, you won't have to relearn anything as you expand your capabilities.
- Since Microchip manufactures both the chips and the assembler, you know you are on their path and not a side road. You won't get left in the dust in the future.
- The examples in Microchip's embedded control handbook and in magazine articles are, for the most part, in Microchip's assembler dialect - so why not go mainstream?
- The error detection feature will be very helpful.

You will need only a small portion of the MPASM assembler's capabilities to get started. An explanation follows:

SOURCE CODE FOR THE ASSEMBLER

Code for PIC16/17's may be created using a simple text editor such as QEdit (designed for programmers), PC Text or EDLIN (included in MS-DOS). Any simple text editor may be used as long as it generates an ASCII file without additional formatting or printer control commands. They tend to confuse the assembler.

Specific information such as which PIC16/17 part the code is written for and where the program starts in the chip's (E)EPROM plus the instructions and labels (mnemonic names for addresses) are typed in a specified format so that the assembler program can find them.

A very simple example will be used to show how all this is done. The program teaches a PIC16C84's port "B" that all eight port lines are output lines (as opposed to input lines) and then asserts the least significant four lines logic 1 (HI) and the most significant four lines logic 0 (LO). Finally, the microcontroller sits in a loop. Definitions for these terms follow.

The source code is easy to lay out using a text editor.

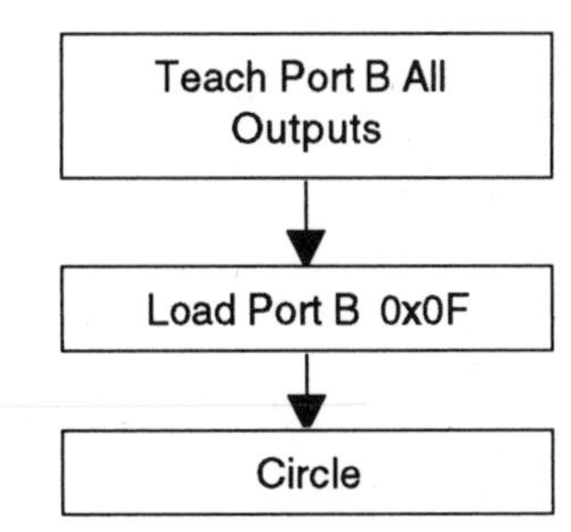


```
;=======PICT9.ASM=============================6/26/96==
        list    p=16c84
        radix   hex
;-----------------------------------------------------
;       cpu equates (memory map)
portb   equ     0x06
;-----------------------------------------------------
        org     0x000
;
start   movlw   0x00     ;load W with 0x00
        tris    portb    ;copy W tristate, port B
;                           outputs
        movlw   0x0f     ;load W with 0x0F
        movwf   portb    ;load port B with contents
;                            of W
circle  goto    circle   ;done
;
        end
;------------------------------------------------------
;at blast time, select:
;       memory unprotected
;       watchdog timer disabled (default is enabled)
;       standard crystal (using 4 MHz osc for test)
;       power-up timer on
;=====================================================
```

Semicolon (;)

The semicolon (;) is a delimiter which tells the assembler to ignore everything following the semicolon on that line of text. If the semicolon is the first character on a line, the whole line is ignored by the assembler. This is useful for putting in your own symbols to divide the text into sections visually (--------- or =======) or for putting in comments or notes to explain and document the program.

On the line which begins with "start" in the example, the information following the semicolon "load w with 0x00" explains what the MOVLW instruction is being used for.

Tabs

The tab capability of the text editor is used to arrange code in columns. The assembler expects to find specific items in three columns. The three columns are used five ways, one way for the header section, a second way for the equate section (defined later), a third way for org statements, a fourth way for the program section, and a fifth way for the end statement.

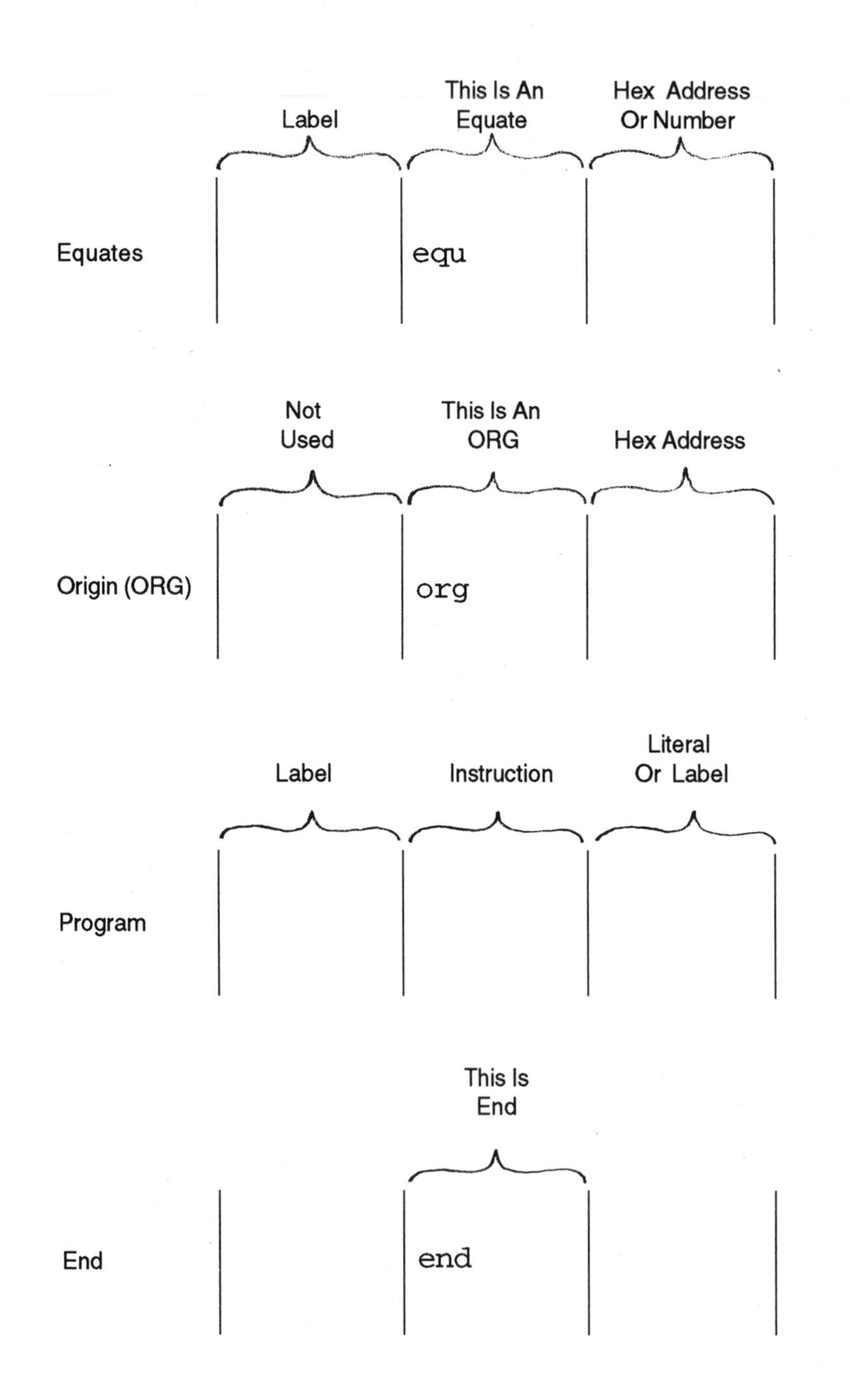

The assembler program "reads" each text file line, ignores it if it begins with a semicolon, and steps through the three columns. If it finds "EQU", "ORG", or "END" in the second column, it treats that as an EQUATE, ORG, or END command. If it finds one of the PIC16C84 program commands in the second column, the assembler creates machine language code in hexadecimal.

Tab position (i.e. 8th column vs 9th or 10th, etc.) is not critical. When the assembler sees a tab (or a space does the same thing) as it moves from left to right along the line, it knows there is a transition from one column to the next.

STYLE

As you look at source code examples (text code), you will see quite a range of styles in dividing up the page, placement of comments, etc. Don't let this confuse you. It is just a matter of differing styles of the individuals who created the programs. It is what's in the three columns not blanked out with the use of ;'s that counts.

```
          list      p=16c84
          radix     hex
portb     equ       0x06
          org       0x000
start     movlw     0x00
          tris      portb
          movlw     0x0f
          movwf     portb
circle    goto      circle
          end
```

HEADERS

The information at the top of the source file is called a header.

```
;======PICT9.ASM===================================6/26/96==
          list        p=16c84
          radix       hex
;----------------------------------------------------------
```

The line

```
          list        p=16c84
```

indicates which PIC16/17 part the program will be placed in. "List" is an assembler directive which has a bunch of meanings depending on what follows it. This is the only way we will use the list directive.

The line

```
          radix       hex
```

indicates that the numbering system is hexadecimal unless otherwise specified in a specific instruction.

For all examples in this book, it is assumed that a header similar to the the one shown above is used.

LABELS

A label is a mnemonic symbolic name assigned to an address. "portb" means port B whose real hexadecimal address is 0x06 in the file register = data memory space in a PIC16C84. Labels are nice because once assigned via an EQUATE statement equating the symbol with a hexadecimal physical address, you no longer have to remember port B is at 0x06. Just refer to it in your program via the label "portb".

Labels are also assigned merely by putting them in the label column of the source file. This is done when the address can be anywhere, i.e. it does not have to be related to something physical such as a port with a specific address. Labels defined in this way are automatically assigned addresses by the assembler. We don't need to know what they are.

The rules for defining labels are:

- All labels must start in the first position in column 1.
- Labels must begin with an alpha character or an underbar.
- Labels may contain alphanumeric characters, the underbar and the question mark.
- Labels may be up to 31 characters long (you will probably run into the next column before you run out of characters).
- Labels are case sensitive by default.

The underbar (_) is useful as a means of separating words as spaces are not allowed. An example is temp_file.

EQUATES

An equate statement may serve to assign a label to a specific address in the PIC16/17 designated by a hexadecimal number.

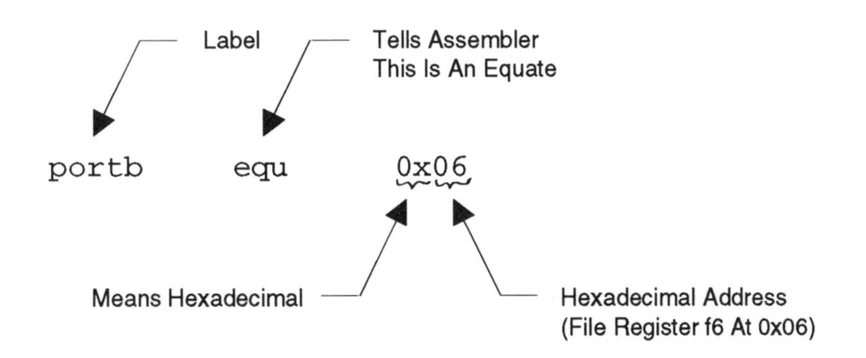

Equates may also be used to assign names to numbers. For example, we can assign the name "min" to bit 1 in a register labeled "flags" with two equates.

```
min         equ         1           ;min = bit 1
flags       equ         0x09        ;flags = file register 0x09
```

In a program, we could have a line with says:

```
bsf        flags,min ;set bit 1 in file register 0x09
```

As the assembler assembles a program, it comes to a label and looks up it's corresponding address or number (not the reverse).

LITERALS

Literals are constants or numbers, usually hexadecimal numbers. Literals are defined using the MOVLW and some logic and arithmetic instructions which are detailed in the programming chapter.

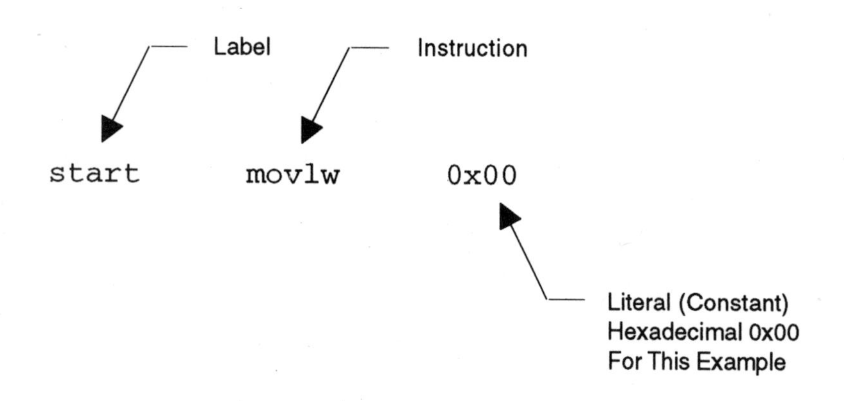

ORG

ORG stands for origin. ORG statements will be used for three purposes in this book.

- ORG defines the address where the program code starts.

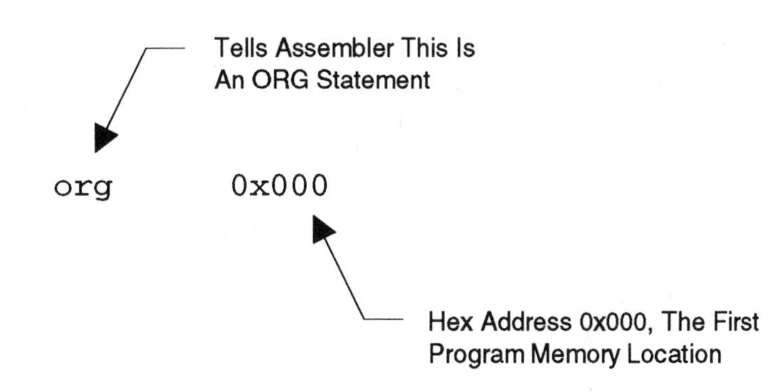

- ORG is used to establish the reset vector for the PIC16C54 (details later).

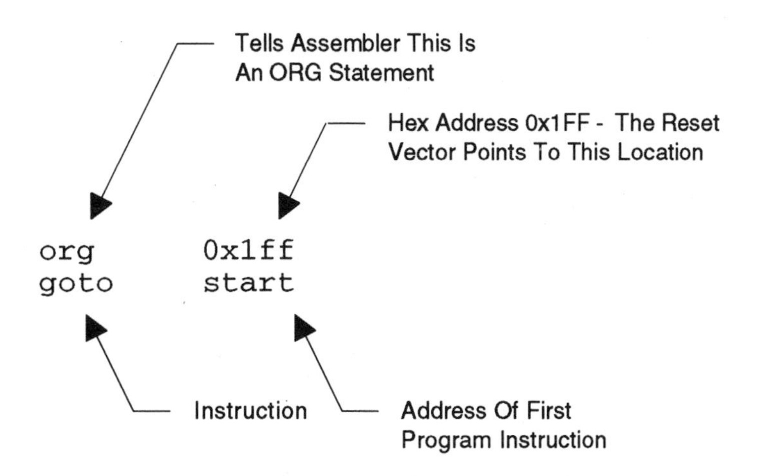

- ORG is used to establish the start of the interrupt service routine for the PIC16C84.

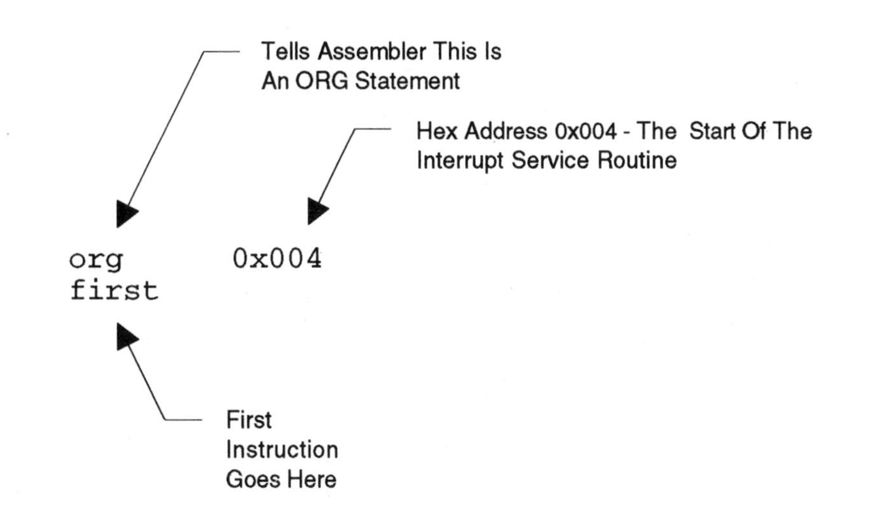

END

An END statement is used to tell the assembler it has reached the end of the program.

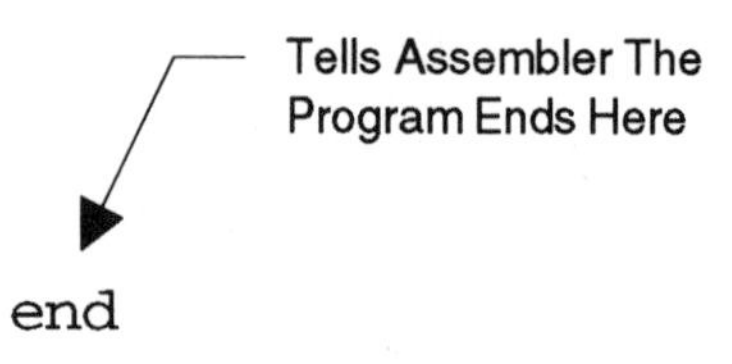

An END statement must be used to end a program.

PROGRAM FORMAT

All program examples in this book are assumed to have the following format even though the header, equates and end are not shown:

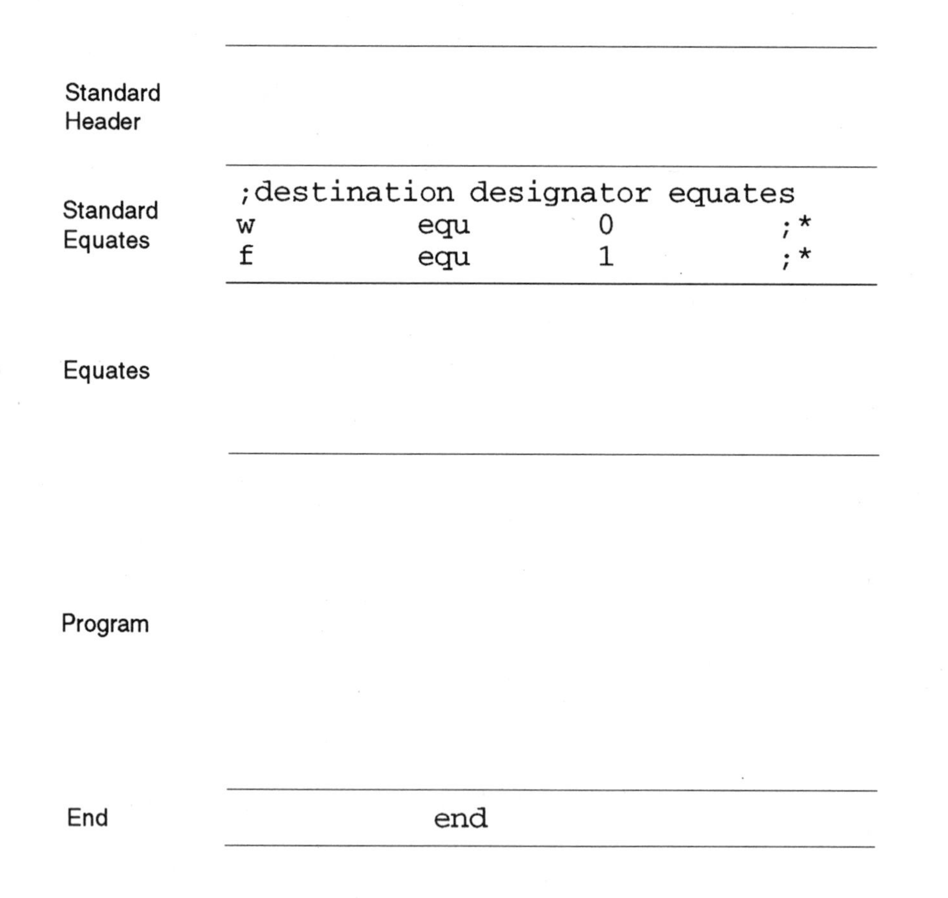

* Explained in destination designator section of programming chapter.

Remember, the first line is always the processor type designation and the last line is always the END statement.

FILES USED BY THE ASSEMBLER

Your program source files created using a text editor must have a file name with the extension .ASM. The file name for the first example in the book is PICT1.ASM. The assembler will only assemble files with the .ASM extension.

FILES CREATED BY THE ASSEMBLER

The assembler will create two files from your source code file which will be of interest to you. They will have file names the same as the source file except for the extensions.

.LST
.HEX

The file whose name has the extension .LST is the assembler-generated listing of your program.

```
MPASM 01.11 Released      PICT9.ASM    6-21-1996   1:54:7                 PAGE  1


LOC  OBJECT CODE       LINE SOURCE TEXT
  VALUE

                       0001 ;=======PICT9.ASM===========================6/26/96==
                       0002          list     p=16c84
                       0003          radix    hex
                       0004 ;-----------------------------------------------------
                       0005 ;        cpu equates (memory map)
  0006                 0006 portb    equ      0x06
                       0007 ;-----------------------------------------------------
                       0008          org      0x000
                       0009 ;
0000 3000              0010 start    movlw    0x00     ;load w with 0x00
0001 0066              0011          tris     portb    ;copy w tristate, port B
                       0012 ;                             outputs
0002 300F              0013          movlw    0x0f     ;load w with 0x0F
0003 0086              0014          movwf    portb    ;load port B with contents
                       0015 ;                             of w
0004 2804              0016 circle   goto     circle   ;done
                       0017 ;
                       0018          end
                       0019 ;------------------------------------------------------
                       0020 ;at blast time, select:
                       0021 ;        memory unprotected
                       0022 ;        watchdog timer disabled (default is enabled)
                       0023 ;        standard crystal (using 4 MHz osc for test)
                       0024 ;        power-up timer on
                       0025 ;======================================================
                       0026
```

```
SYMBOL TABLE

LABEL                             VALUE

__16C84                            0001
circle                             0004
portb                              0006
start                              0000


MEMORY USAGE MAP ('X' = Used,  '-' = Unused)

0000 : XXXXX----------- ---------------- ---------------- ----------------
0040 : ---------------- ---------------- ---------------- ----------------

All other memory blocks unused.


Errors    :     0
Warnings  :     0
Messages  :     0
```

The file whose name has the extension .HEX is the hexadecimal object code which will be used by the programmer to program the PIC16/17 chip.

```
:0800000000306600F3086009D
:020008000428CA
:00000001FF
```

PREVENTING SOME GOTCHAS

More Than One Way To Skin A Cat

There may be two or or more ways to accomplish creating various aspects of a program with an assembler. Specifying the microcontroller type can be handled using a "list" instruction (as done in this book) or via an assembler command line option (not used in this book). As an example, the approach taken here is to use one way consistently. Hopefully, it is the easiest way.

UPPER/lower Case

Use of upper vs lower case letters seems to be very inconsistent in the program listings I have studied. Sometimes it is significant and sometimes it is not (labels are case sensitive for example).

Instruction names and port designations (A vs B) are upper case in the text in this book because that format seems to be the norm. All text in the assembler program listings except comments are lower case to make them less confusing to read and so that case sensitivity is not an issue.

FIRING UP THE MPASM ASSEMBLER AND PICSTART

Microchip supplies two disks in their PICSTART(tm) package. One is the PICSTART disk and the other contains MPASM. The PICSTART disk includes an instruction file which is triggered by typing "INSTALL." Part way through the PICSTART disk procedure, a message on the screen asks for the MPASM disk (by another name) and it gets installed, too.

Sooo......here is the procedure:

1) Create a directory called \PIC. MPASM and the PICSTART driver will be installed in this directory.

2) Insert the PICSTART disk in drive A. My disk is labeled "PICSTART 16B1 Software Tools MPS16B.EXE."

3) Type A: , return

Dialogue boxes will appear on the screen during the installation.

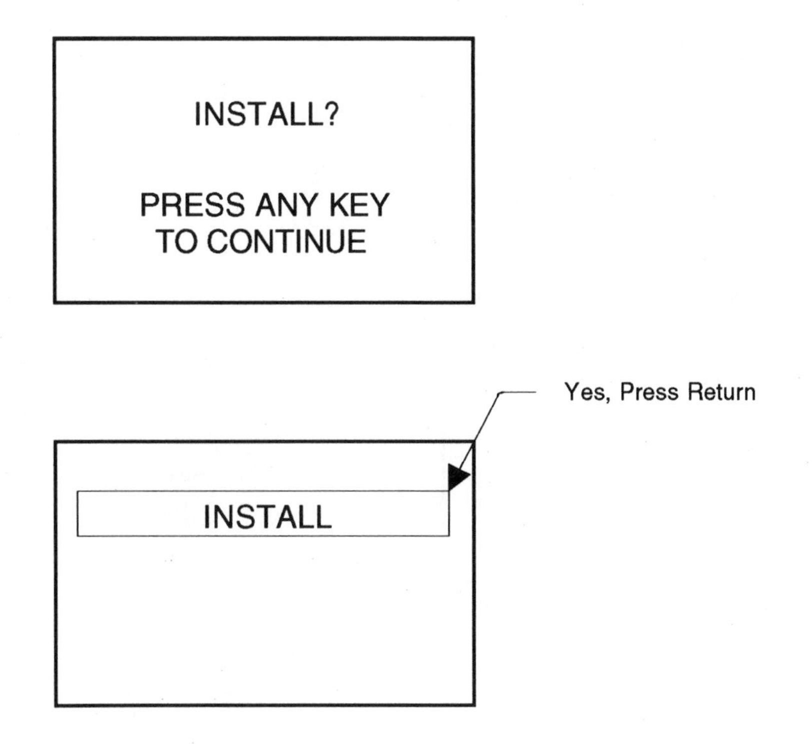

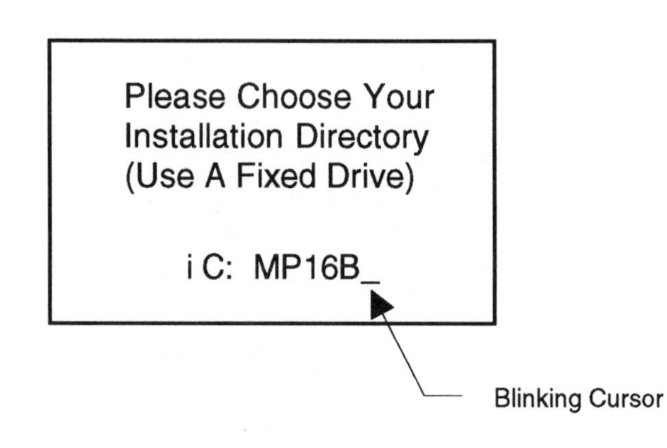

4) Backspace to the back slash to delete "MPS16B" and type "PIC". The result is:

[C:\PIC , return

Since you will have to type the directory name many times, I think you will find it convenient to type "PIC" instead of "MPS16B."

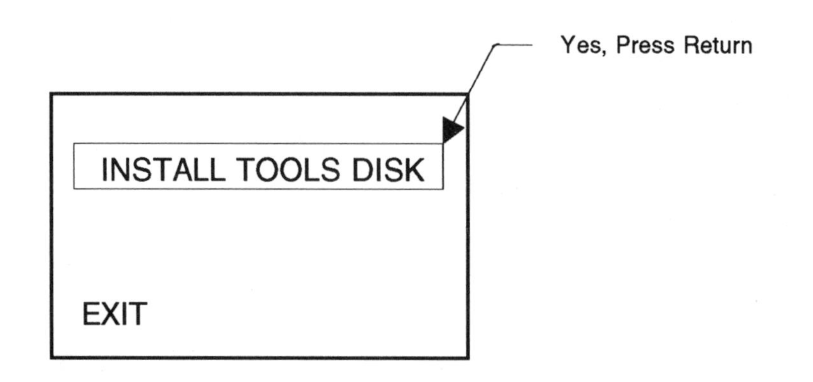

5) At the prompt, insert the tools disk (MPASM disk) in drive A: and press return.

My tools disk is actually labeled "PIC16/17 Tools".

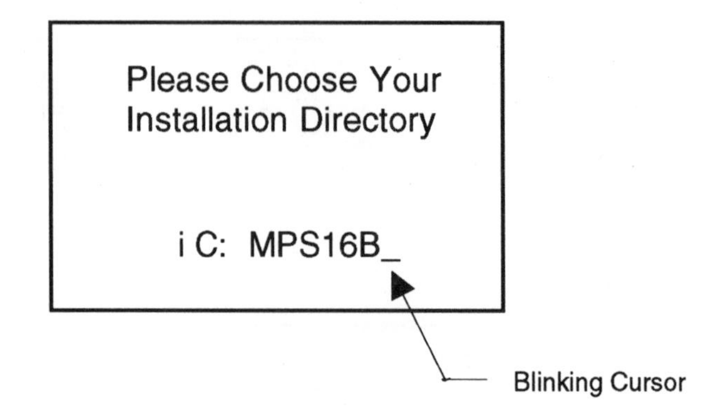

6) Backspace to the backslash to delete "MPS16B" and type "PIC". The result is:

[C:\PIC , return

The installation program copies the MPASM stuff onto the hard drive in the directory called \PIC.

INSTALLATION
COMPLETE

THANKS

MPASM and the PICSTART driver are now ready to be used.

HOW TO ASSEMBLE A PROGRAM

Just to get the feel of how all this stuff works, you may want to type the example program on page 17 using your text editor and assemble it.

- Type it exactly as shown so it will work. There will be plenty of opportunities to make mistakes of your own later.
- Note that if the text is too wide, it will flow into the next line. The overflow will look like a label to the assembler (example: ======. = is an illegal character for a label).
- Name the file PICT1.ASM.
- Assemble using MPASM. PICT1.ASM and MPASM must be in the same directory.
- Type MPASM.
- Press return
- Wait.
- A list of files, etc. will appear.

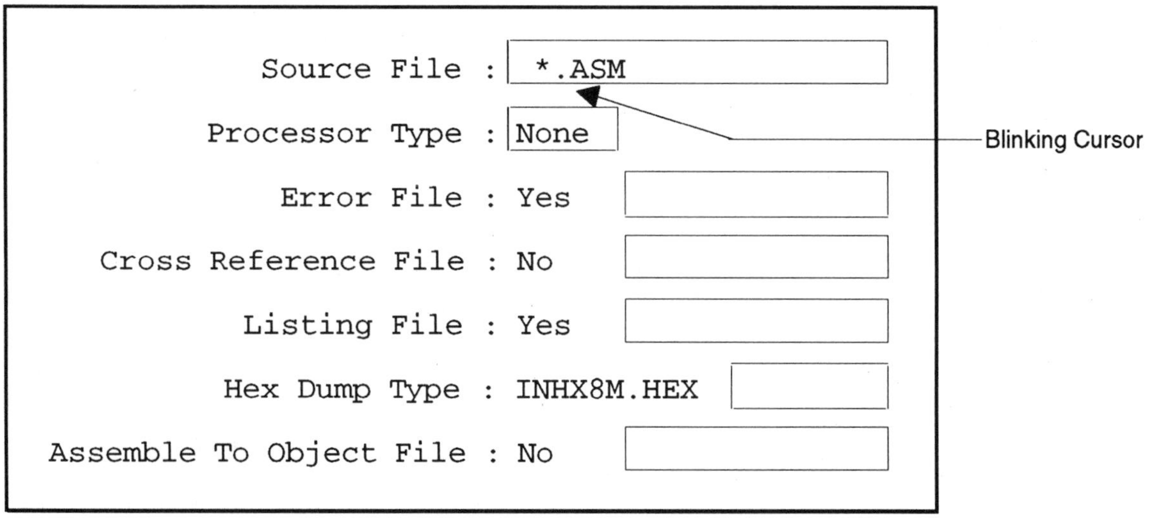

```
                Source File : *.ASM
             Processor Type : None
                 Error File : Yes
       Cross Reference File : No
               Listing File : Yes
              Hex Dump Type : INHX8M.HEX
   Assemble To Object File : No
```

Blinking Cursor

Note that as you work with MPASM that many of the menu selections may be toggled using the return key.

- Press return.
- All .ASM files are shown.
- Use arrow keys to select (highlight) PICT1.ASM.
- Press return.
- Observe that PICT1.ASM has been entered in the appropriate places automatically.

```
               Source File : PICT1.ASM
            Processor Type : None
                Error File : Yes PICT1.ERR
      Cross Reference File : No
              Listing File : Yes PICT1.LST
             Hex Dump Type : INHX8M PICT1.HEX
 Assemble To Object File   : No
```

- Arrow down - highlight processor type.
- Press return.

Processor Type : 16C84

- Arrow down - highlight error file.

```
Error File : Yes
```

- Press return.

```
Error File : No
```

• The result is:

```
          Source File : PICT1.ASM
       Processor Type : 16C84
           Error File : No
 Cross Reference File : No
         Listing File : Yes PICT1.LST
        Hex Dump Type : INHX8M PICT1.HEX
Assemble To Object File : No
```

• Press F10 to assemble.
• When assembly is complete, a list will appear on-screen showing how many errors there are (hopefully 0) and where they are.
• Press any key to get out.
• When the assembler has finished, use the text editor to open the file PICT1.LST to see what the assembler created. Errors will be noted if there are any.
• If you make mistakes, consider them as serving to create an opportunity to find out what the error codes mean. They may be corrected by opening the .ASM text file using the text editor. Then repeat the assembly process.

HOW TO USE THE PICSTART PROGRAMMER

- Connect the PICSTART to the computer via the serial cable (see manual for detailed instructions).
- Plug in the wall transformer and connect the cable to the board. The green LED will light indicating that the PICSTART is powered.

Do Not Power Up PICSTART With Device In Socket

Never Insert Or Remove Device With Yellow = Active LED On

- Type MPS16B, return (assuming you have a PICSTART-16B).
- Click mouse on Device Edit.

16C54 will be selected.
Click on ( ) to select another device if desired.

- Click on OK.
- Click on Fuse Edit.

Dialogue box

( ) LP [X] Watchdog timer on
(•) RC [ ] Code protect on
( ) XT [X] Power-Up timer on
( ) HS

- Select clock oscillator type by clicking on ( ).
- Select watchdog timer off by clicking on [ ].
- Code protect off (default).
- Power-up timer on (default).
- Click OK.

Note that you can see the "config" selections at the upper right hand corner of the screen to double-check your selections.

- Click on File menu.
- Click Open.

Dialogue box.

- Click on file name.
- Click on Open.

Your hex file will be on the screen.

- Click on Program menu.
- Click Program.

VARIOUS MESSAGES
ENDING WITH:

"PROGRAMMING
COMPLETE!"

- Alt-X to exit.

WRITING PROGRAMS

PROGRAMMING CONCEPTS

The PIC16C84 microcontroller will respond to a series of coded instructions stored in program memory.

When a designer (which may be you) thinks of something which he or she would like to control, the natural thing to do is to think through the control process in logical steps. Creating a flow chart is a good way to visualize these steps. The control process might consist of sensing outside world events such as light vs. darkness, cows passing through a gate, temperature, a key stroke etc., testing the data which has come from the sensors followed by taking one of two possible program paths (branching) based on the test results, and controlling some outside device such as a digital display, indicator light, motor, heater, etc.

Instructions for repetitive operations can be repeated in long strings, but that wastes valuable memory space. It may even result in a program which is too long to fit in the available memory space. Instead, one sequence of instructions can be used over and over in a loop and the microcontroller goes 'round and 'round until something forces it to stop.

- Loops can go around forever or until the plug is pulled or an anvil is dropped on the microcontroller chip.
- Or until a counter counts up to a predetermined number or down to 0.
- Or until a test result says to move on.

An example of the use of a loop is a program used to read an input port and display the data received (or the status of the lines) at an output port.

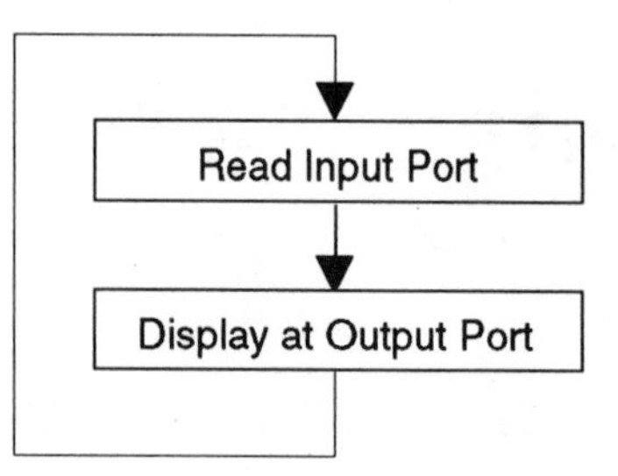

Loops are useful when it is necessary to perform an operation a certain number of times (n).

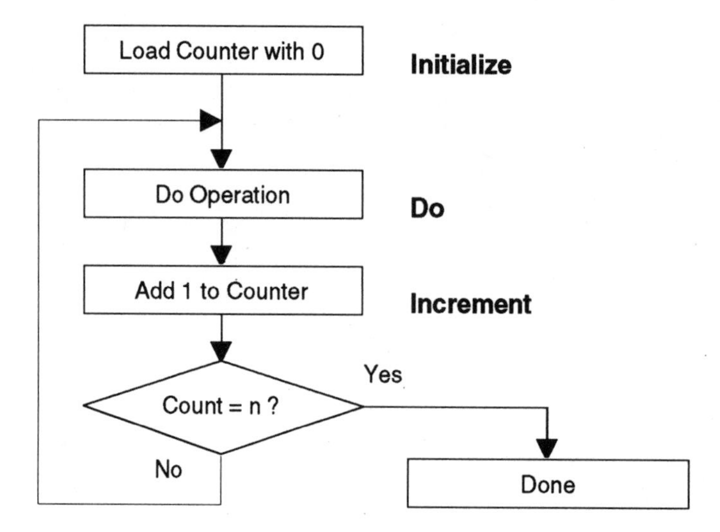

The use of a loop prevents having to write the code n times and the requirement for memory space to store it.

Another technique for keeping programs short and manageable is the use of subroutines. If the same task is to be performed in two or more places in a program, the code can be written once and stored in one location. When the points in the program are reached where the subroutine is to be used (called), a CALL instruction is used. The last instruction in a subroutine is always a return from subroutine (RETURN or RETLW) instruction. The program then continues where it left off.

An important concept to keep in mind when using micros is that they can only do one thing at a time, one very simple thing. They execute many very simple instructions and they do it blindingly fast. The PIC16C84 executes roughly a million instructions every second with a 4MHz clock oscillator.

In situations where events of interest to the micro occur only once in a while, perhaps randomly, it may be desirable to use a micro feature called the interrupt. In simplified form, if an event occurs in the outside world which demands the micro's attention, the sensor monitoring the event can be wired to direct a signal (pulse) to an interrupt line (pin) on the microcontroller. When the signal arrives, the microcontroller drops what it was doing (after finishing the instruction it was executing), and then jumps to a special program called an interrupt service routine. The purpose of the routine is to do whatever the designer/ programmer thinks is required when the outside event occurs. The microcontroller saves processor status and register information for use when the micro goes back to what it was doing.

Micros do only one thing at a time, but they can be instructed to drop one task, take care of another, and resume the original task.

Interrupts are a special area closely tied to the hardware used to make their occurance known, so they will be discussed in detail in the interrupt section of this manual.

Finally another basic concept is that there is no way to write directly to a memory location or to an I/O Port. Data must be put into the W register and then the W register contents must be stored in the final location. A MOVLW or MOVF instruction loads the W register. A MOVWF instruction stores a copy of the W register contents in a data memory location (file register) leaving the contents of the W register unaltered.

PROGRAMMING EXAMPLES

The best way to learn how micros are used is to think of applications and write programs to implement them. In this section, we will start out with very simple programs to demonstrate the concepts we have just discussed. By the time you have finished this section, you will be able to think in micro terms and will see micro applications in your work or hobbies. You will also be able to visualize the methods for implementing micro solutions.

Simple programs used as examples will illustrate the use of the various types of instructions.

INSTRUCTION SET

The PIC16C84 microcontroller officially has 35 instructions in its instruction set. We will temporarily use two obsolete instructions because it makes life much simpler (total 37 instructions to get started). The instruction set is presented categorized by type of operation.

The use of k, f, d, etc. will be explained later.

PIC16/17 instruction words used by the chip itself contain both the binary code for the instruction and the binary address. For the PIC16C84, instruction words are 14 bits wide. How these words are constructed is described in the Microchip data book. I have found that you simply won't need to know how this is done. Getting the assembly language instructions and designators correct should be the focus.

Move or Define Data

MOVLW	k	Loads W with literal.
MOVF	f,d	Moves copy of selected register contents into W or f.
MOVWF	f	Moves copy of W contents into selected register.

Change Register Contents

CLRF	f	Clears selected register to 0.
CLRW		Clears W register to 0.
COMF	f,d	Complements selected register contents.. All 1's to 0's, all 0's to 1's Result in W or f.
DECF	f,d	Decrements selected register. Decrementing when contents of register is 0x00 results in 0xFF. Result in W or f.
INCF	f,d	Increments selected register. Incrementing when contents of register is 0xFF results in 0x00. Result in W or f.
BCF	f,b	Clears selected bit in selected register to 0.
BSF	f,b	Sets selected bit in selected register to 1.
RLF	f,d	Rotates bits in selected register one position to the left. Bits rotate through carry flag. Result in W or f.
RRF	f,d	Rotates bits in selected register one position to the right. Bits rotate through carry flag. Result in W or f.

SWAPF	f,d	Exchanges MS and LS nybbles of selected register. Result in W or F.

Control Program Flow

GOTO	k	Go to specified address.
CALL	k	Call subroutine at specified starting address.
RETURN		Return from subroutine.
RETLW	k	Return from subroutine. Loads W with literal.
RETFIE		Return from interrupt.
BTFSC	f,b	Tests specified bit in specified register. Skips the next instruction if bit tested is clear (0).
BTFSS	f,b	Tests specified bit in specified register. Skips the next instruction if bit tested is set (1).
DECFSZ	f,d	Decrements specified register. Skips next instruction if registe contents = 0. Destination W or f.
INCFSZ	f,d	Increments specified register. Skips next instruction if register contents = 0 Destination W or f.

Nothing

NOP		Do nothing for one instruction cycle (time delay, save room for future code mods, debugging = break).

Control Microcontroller

CLRWDT		Clear watchdog timer (reset to zero). Also resets the prescaler of the watchdog timer. Status bits TO and PD are set.
OPTION		W contents (bit pattern) sent to option register to control prescaler ratio, real time clock trigger edge, and real time counter/clock source.
SLEEP		Puts microcontroller to sleep to reduce power consumption. Wakeup via reset, watchdog timer, or external real time input.
TRIS	f	W bit pattern determines port line input vs. output on line-by-line basis for selected port.

Logic

ANDLW	k	AND's contents of W with literal (mask) contained in instruction. Result in W.
ANDWF	f,d	AND's contents of W with contents of selected register. Result in W or f.
IORLW	k	OR's contents of W with literal (mask) contained in instruction. Result in W.
IORWF	f,d	OR's contents of W with contents of selected register Result in W or f.
XORLW	k	XOR's contents of W with literal (mask) contained in instruction. Result in W.
XORWF	f,d	XOR's contents of W with contents of selected register. Result in W or f.

Arithmetic

ADDWF	f,d	Adds contents of W to contents of selected register. Results in W or f.
ADDLW	k	Add literal to W.
SUBLW	k	Subtract W from literal. Result in W. (NOT subtract literal from W as name indicates!).
SUBWF	f,d	Subtracts contents of W from contents of selected register by 2's complement arithmetic. Results in W or f.

INSTRUCTION FORMAT FOR ASSEMBLER

First, some definitions:

f = file register
d = destination
0 = W register
1 = file register
k = constant (literal) (literal instructions)
k = address label (call or goto instructions)
b = bit designator (bit-oriented instructions)
b = binary (literal instructions)
d = decimal (literal instructions)

There are four categories of instructions when it comes to formatting instructions for the assembler.

Byte-Oriented Instructions

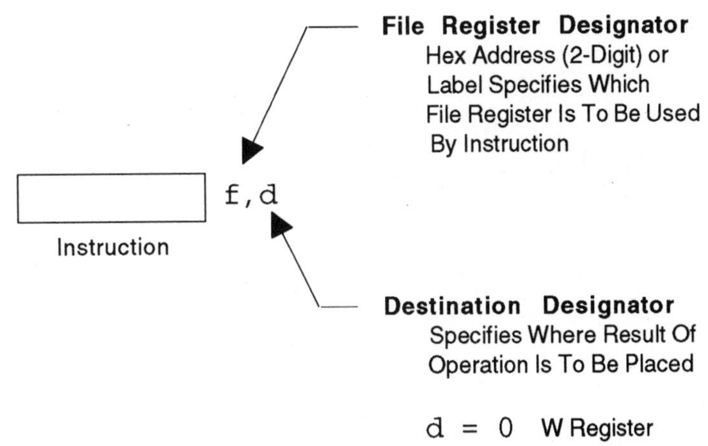

Bit-Oriented Instructions

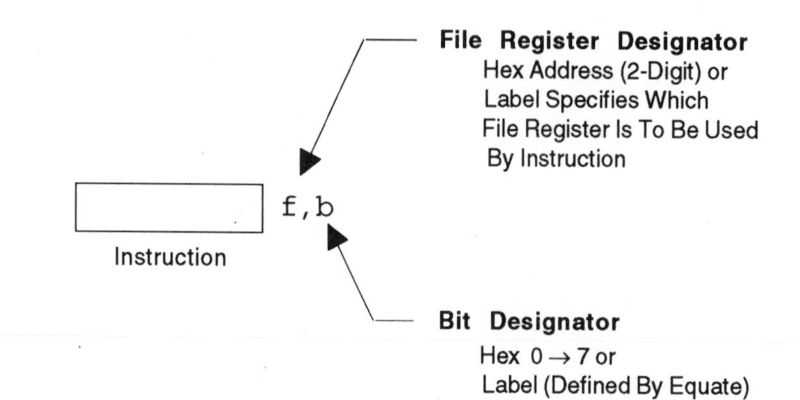

Literal Instructions

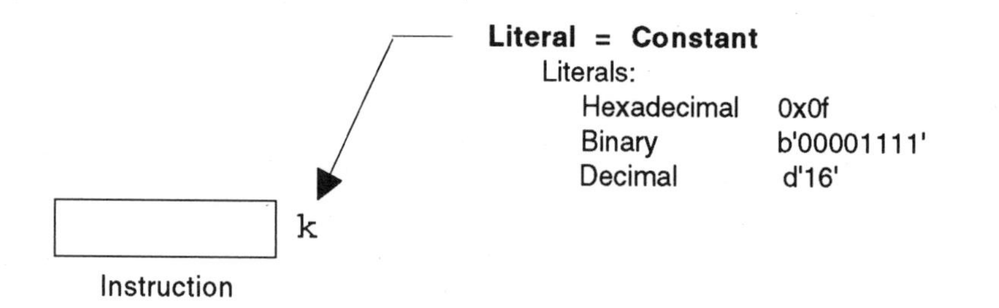

Control Instructions (CALL and GOTO)

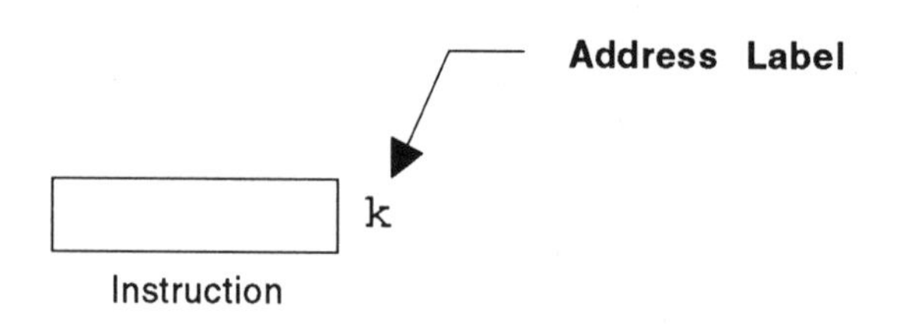

DESTINATION DESIGNATOR (d)

Byte-oriented instructions are accompanied by a designation designator referred to as "d".

d = 0 Destination is W register (result in W).
d = 1 Destination is a file register (result in f).

Trying to remember if d should be 0 or 1 is a nuisance. To avoid having to deal with it, all programming examples in this book from this point forward are assumed to have the following equates ahead of them:

```
w           equ         0
f           equ         1
```

Examples:

```
            decf        counter,f       ;f means result in file
;                                             register
            movf        counter,w       ;w means result in W register
```

Note that if the contents of a file register are operated on and the destination of the result is W, the contents of the file register remain unchanged.

```
            incf        value,w         ;result to W
            rrf         value,w         ;result to W
```

In both cases, the result goes to W and the contents of the file register labeled "value" are unchanged.

HEXADECIMAL NUMBERS vs. MPASM ASSEMBLER

The use of hexadecimal numbers with PIC16/17's is full of inconsistencies! You will see this when you look through program listings from other sources. For example port B may be equated to the file address hexadecimal 06 in the following ways:

```
portb       equ         6
                        06
                        06h
                        h'06'
                        0x06
```

The MOVLW instruction is used to load the W register with hexadecimal literals as follows:

```
            movlw       00
                        00h
                        h'00'
                        0x00
                        0f
                        ff          won't work
                        ffh         won't work
                        h'ff'
                        0xff
```

If 00 and 0f work, why doesn't ff work? It looks like the same form to me. The important thing is to be aware of the inconsistencies and use a format that always works.

The first character in the literal expression must be "0" or "h" for the assembler to work.

Sooooo .... to make things manageable we will settle on a standard/uniform way of doing things for the examples in this book.

Single hex digits by themselves will be used for:

Equates	bits 0 → 7
Instructions	bit designator b 0 → 7

Hex addresses will be in the following form:

File registers = data memory	0xXX
Program memory	0xXXX

Hex numbers in literal instructions will be in the following form:

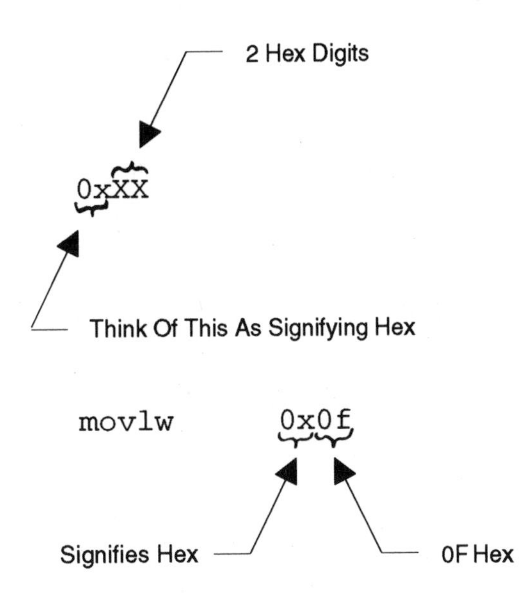

The programs in this book are written using these conventions.

All references to hexadecimal numbers in the text of the book will use the 0x notation.

BINARY AND DECIMAL NUMBERS vs. MPASM ASSEMBLER

Binary and decimal literals may be written as shown:

```
movlw       b'00001111'       ;binary
movlw       d'16'             ;decimal
```

Note that the ' is the apostrophe on the same key as " on the keyboard. Some listings I have seen appear to have the literal bracketed in ` '. Only ' ' works with MPASM.

ADDRESSING MODES

- IMMEDIATE - using "literal" instruction which loads data from program memory into W register.
- DIRECT - specific address.
- INDIRECT - using file select register (FSR) (f4).
- RELATIVE

  via bit test and skip (2 addresses away).
  via loading computed addresses into the program counter. Could be out into the middle of a bunch of RETLW's (lookup table) or GOTO's (jump table).

Immediate Addressing

The literal instruction has the data (literal) built into the instruction word.

Direct Addressing

The direct addressing mode is straightforward. The address is specified following the instruction.

```
        clrf        temp        ;clear file labeled temp
```

Indirect Addressing

The first file register f0 (indirect address or indirect file, INDF) at address 0x00 is not physically implemented (it ain't there!). Using 0x00 as an address actually calls for the contents of the file select register f4 (FSR at 0x04) to be used as the address of a file register. The indirect address register is bogus. Using its address tells the PIC16/17 that what you really want is the address pointed to by the FSR. Weird, but that's how it works.

If you want to take off relative to a known file register address, add an offset or index to that address, store the result in the FSR and use 0x00 as the file register designator for the instruction which takes you to the resulting address. This is like "indexed addressing" with other microcontrollers. Since there are only 36 general-purpose file registers in a PIC16C84, you won't be indexing far.

```
; assume address is in file labeled "hold"

        movf        hold,w          ;contents of hold (address) to W
        movwf       fsr             ;W to fsr
        movf        0x00,w          ;load data at address into W
        continue
```

To review, using the f0 address cause the contents of the FSR to be used as the address of the file register. F0 is useful as an address pointer.

A practical application for this might be sending characters from 16 file registers 0x20 to 0x2F to an LCD display.

```
        movlw       0x20        ;initialize pointer
        movwf       fsr,f
next    movf        indf,w      ;get data pointed to
        call        send        ;sub sends byte to LCD
        incf        fsr         ;inc pointer
        btfss       fsr,4       ;done ?
        goto        next        ;no, again
```

To implement this, one could load the "display RAM" in advance using MOVLW instructions.

Relative Addressing

Relative addressing for the PIC16/17 involves altering the contents of the program counter. This constitutes a computed jump from wherever the program counter is to some address relative to that point.

For GOTO and CALL instructions on the PIC16C84, this is handled automatically because the address is included in the instruction.

The 5 high bits of the PC are loaded from PC latch high (PCLATH) by any instruction which writes to the PC. All computed jumps must be preceded by being sure that PCLATH contains the correct bits. Since we are avoiding program memory paging for now, we are limited to using the first 256 locations of program memory for computed jumps.

So, if you want to take off from where the PC is, add the offset or index to the PC and execution will continue at the resulting address. Use the contents of a counter as an offset or index to jump to a code chunk or to a GOTO.

Use Of The RETLW Instruction For Accessing Tables of Data Via Relative Addressing

The RETLW instruction causes the W register to be loaded with the literal value following the RETLW instruction. This is useful for implementing data tables in program memory. Limited amounts of data can be stored in program memory and accessed this way.

As an example, let's say we have an application where a counter (file register) is incremented (0 to some number = 9 or less) and we want to display the result on a 7-segment LED display. Seven segment codes are stored as a table in program memory with each code attached to a RETLW instruction (coupled with RETLW instructions). This is a way to store 8-bit data in 14-bit program memory and gain access to it.

The code which must access the table calls the subroutine which contains the table. The counter is labeled "count". After the counter has been incremented the last time, the counter contents are loaded in W and the subroutine is called. The first instruction in the subroutine adds the contents of W to the program counter (used as an index or offset). Execution jumps to a RETLW instruction containing the 7-segment code corresponding to the counter contents. Execution returns to the main program and the W register now contains the 7-segment code. The code is then sent to the 7-segment display via port B.

The details for this example follow at the end of the programming chapter.

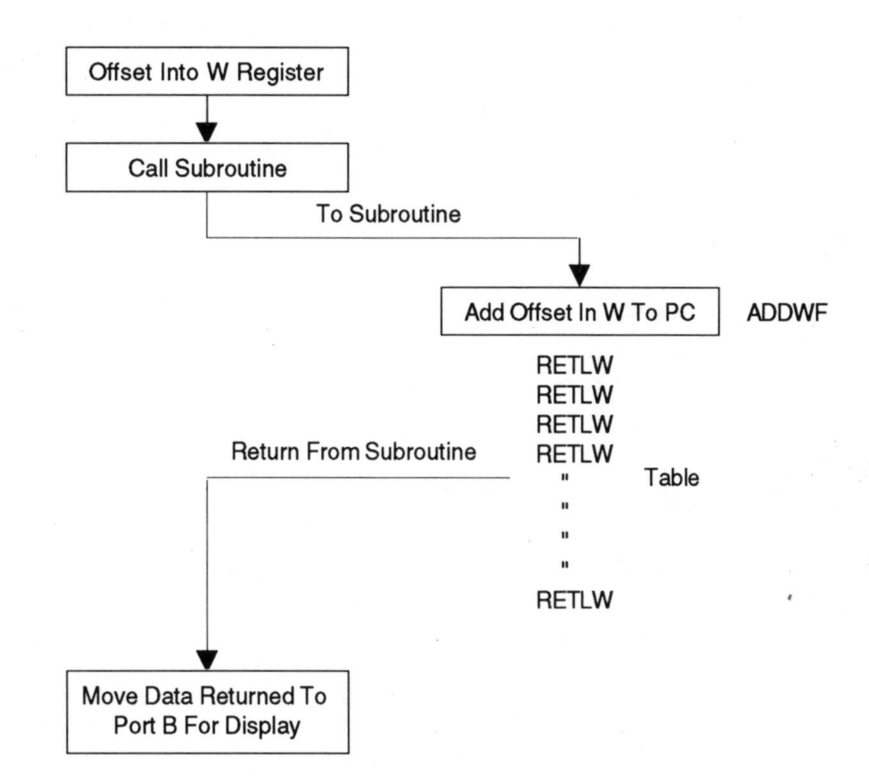


```
pc          equ       0x02        ;program counter
count       equ       0x0C        ;counter


            movf      count,w     ;counter contents to w
            call      segments    ;call sub
            movwf     portb       ;display results


segments    addwf     pc,f        ;add offset to pgm ctr
            retlw     3f          ;0 seven-segment
            retlw     06          ;1
            retlw     53          ;2
            retlw     4f          ;3
            retlw     66          ;4
            retlw     6d          ;5
            retlw     7d          ;6
            retlw     07          ;7
            retlw     7f          ;8
            retlw     6f          ;9
```

USING THE PORTS

Port Data Direction

Port data direction (input vs. output) is controlled using a special register called the tristate register. Like the option register, we will think of the tristate (TRIS) register as a "buried" register, not part of the file register address space and having no address. We will use the TRIS instruction to write to it. This is the second of two obsolete PIC16C84 instructions which we will use because it makes life more manageable. File register bank switching and the use of Microchip's intended methods will come later.

The TRIS and OPTION methods are the only ones available for the PIC16C54, so learning them now is part of the backward compatibility plan.

The TRIS register has 8 bits.

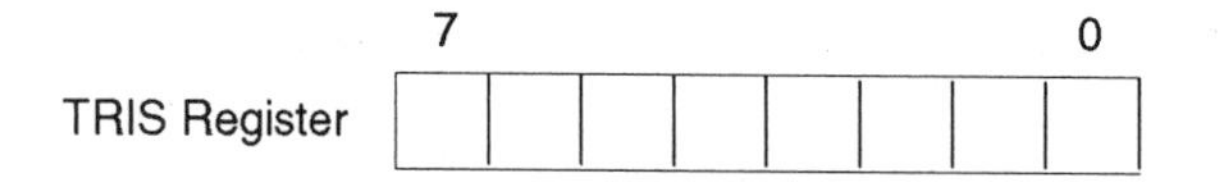

Loading a "0" in a TRIS register bit makes the corresponding port bit an output. Conversely, a "1" results in an input.

```
          movlw     b'00001111'  ;teach port B
          tris      portb        ;bits7,6,5,4 outputs
;                                 bits 3,2,1,0 inputs
```

The TRIS instruction transfers the contents of the W register into the TRIS register which selects port data direction.

Port Read/Write

An input port may be read as a whole by using the MOVF portX, W instruction. Data is valid as of the instant the port is read. An individual port line may be read using the BTFSC or BTFSS instructions.

The bit-oriented instructions BCF and BSF read the whole port contents into the W register, change a bit, and write the contents of the W register back to the port. This is called read-modify-write.

The port is read at the beginning of an instruction cycle and written to at the end of the instruction cycle. If a read immediately follows a write to the same port, problems may occur. Time should be allowed for the port to stabilize. Inserting a NOP instruction between the write and read should prevent this problem from occurring.

Doing two consecutive writes to a port may cause the same problem to occur. Again, inserting a NOP should prevent trouble.

FLAGS

A flag is a one-bit register which is set (1) or cleared (0) by execution of one or more types of instructions. The setting or clearing generally takes place automatically. After execution of an instruction, the affected flag may be tested to see if it was set or cleared. The path taken by the program depends on the status of the flag being tested.

There are three flags in the PIC16C84. We will be interested in two of them (Z and C). For now, it is sufficient to know that these flags exist in the status word register f3. Their significance will be discussed as the applications for them arise.

```
-----------------------
        FLAGS
-----------------------
  Z        Zero
  C        Carry
```

SIMPLE DATA TRANSFERS

This program moves data from port B (file register f6 at address 0x06) to a file register labeled "hold".

```
movf     portb,w    ;move contents of port B to W reg
movwf    hold       ;move contents of W register to
                    ;file labeled hold
```

This requires moving data via the W register. There is no direct way to move data from one file to another.

A data transfer is used to send a bit pattern to an output port. Let's arbitrarily choose the following bit pattern and send it to the parallel output port B located at 0x06 which we will assume drives 8 LED's.

```
Bit Pattern        0000 1111
Hexadecimal        0x0  0xF
```

```
movlw       0x00        ;load W with 0x00
tris        portb       ;teach port B outputs
movlw       0x0f        ;load W with 0x0F
movwf       portb       ;load port B with contents of W
```

To set an output port to all 0's as part of a program, you can use the CLRF instruction which clears the selected register to 0x00.

```
clrf        portb
```

Input ports are read using data transfers. DIP switches are assumed to be located at port A (LO 4 bits)and LED's at port B.

```
;=======PICT2.ASM============================5/19/96==
;data transfer demo
;-----------------------------------------------------
        list    p=16c84
        radix   hex
;-----------------------------------------------------
;       destination designator equates
w       equ     0
f       equ     1
;-----------------------------------------------------
;       cpu equates (memory map)
porta   equ     0x05
portb   equ     0x06
;-----------------------------------------------------
        org     0x000
start   movlw   0xff    ;load w with 0xFF
        tris    porta   ;copy w tristate, port A
;                          inputs
        movlw   0x00    ;load w with 0x00
        tris    portb   ;copy w tristate, port B
;                          outputs
        movf    porta,w ;read port A, result in W
        movwf   portb   ;LO nybble display at port B
circle  goto    circle  ;circle
;
        end
;-----------------------------------------------------
```

```
;at blast time, select:
;       memory unprotected
;       watchdog timer disabled (default is enabled)
;       standard crystal (using 4 MHz osc for test)
;       power-up timer on
;=======================================================
```

Set the switches at the input port & run the program. Compare the resulting bit pattern at the display with the pattern you set in. Change the input pattern and rerun the program.

LOOP - ENDLESS

Above we used an endless loop to fake a halt (the PIC16C84 does not have a halt instruction). The GOTO (go to) instruction goes to the location labeled "circle" which is the address of the GOTO instruction. The PIC16C84 sits in an endless loop (circle).

Next we will use the GOTO instruction to make a more useful loop. The program will read the input port A and display the 4 bits of port A continuously at the lowest 4 bits of port B. This time, the program repeats over and over by jumping back to "start" each trip through the loop. The code is the same as before except for the address of the GOTO instruction.

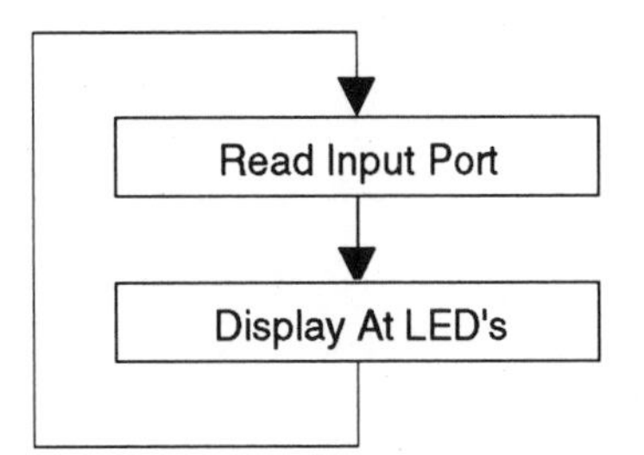

The program is the same as above except for the line beginning with the label "circle" is replaced with:

```
        goto        start       ;again
```

Now vary the switch settings while the program is running.

This loop will run forever unless you press the reset switch or pull the plug (sometimes called "absolute reset").

LOOP WITH COUNTER

It is often necessary to perform some operation a specified number of times. To do this, a file register may be set aside to be used as a counter. Each time the operation is performed, a one is added to the counter. This is called incrementing the counter and the instruction is INCF. When the number in the counter becomes equal to the number of times the operation is to be performed, the program can stop or go on to something else. A simple example will illustrate this concept.

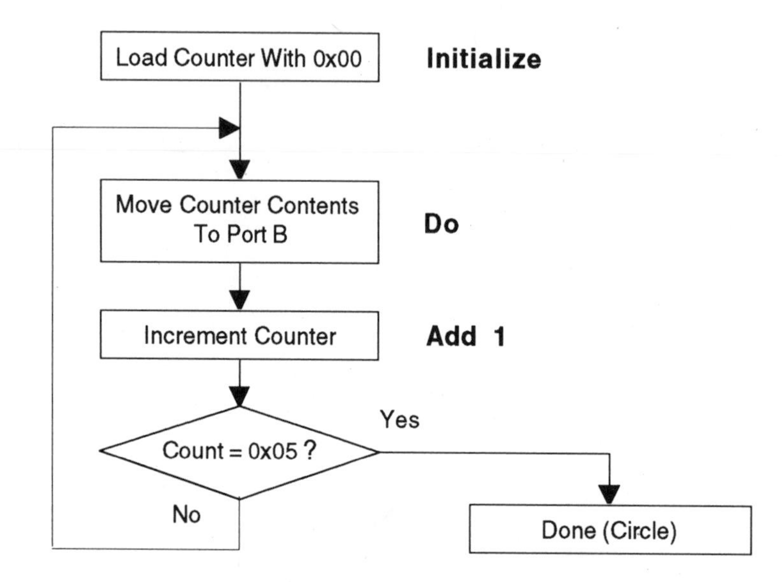

We will use file register f7 (general purpose file register = RAM) labeled "count" as the counter. Storing the counter contents at port B is just something for the program to do. We will find more exciting things to do later in this section.

```
;=======PICT3.ASM===========================5/22/96==
;loop with counter demo
;       count up
;----------------------------------------------------
        list    p=16c84
        radix   hex
;----------------------------------------------------
;       destination designator equates
w       equ     0
f       equ     1
;----------------------------------------------------
;       cpu equates (memory map)
status  equ     0x03
portb   equ     0x06
count   equ     0x0c
;----------------------------------------------------
        org     0x000
start   movlw   0x00     ;load w with 0x00
```

```
        tris     portb    ;copy w tristate, port B
;                              outputs
        clrf     portb    ;outputs LO
        clrf     count    ;clear counter
again   movf     count,w  ;counter contents to W
        movwf    portb    ;counter contents to port B
        incf     count,f  ;inc count, result in count
        movlw    0x05     ;load w with 0x05
        subwf    count,w  ;subtract 0x05 from counter,
        ;                      result in W register
        btfss    status,2;test Z flag
        goto     again    ;not = 0x05
circle  goto     circle   ;= 0x05, done
;
        end
;------------------------------------------------------
;at blast time, select:
;       memory unprotected
;       watchdog timer disabled (default is enabled)
;       standard crystal (using 4 MHz osc for test)
;       power-up timer on
;======================================================
```

Port B will indicate 0x04 in binary (00000100).

This program illustrates a very important concept, the power of microcontrollers to make decisions. A comparison is made using the MOVLW and SUBWF instructions to see if two things are the same. If they are, the program stops. If not, the program loop continues until the counter hits the number we have chosen. The subtraction and bit test instructions (comparison) determine the flow or path of the program.

Load the program and run it to see what happens. It will be finished in a few microseconds. Port B should contain 0x04. You may want to try other numbers in the counter.

You may find it easier to load the counter with the number of times an operation is to be performed and decrement the counter using the DECFSZ instruction. When the count falls to zero, the next instruction (a GOTO) is skipped sending execution to the code following.

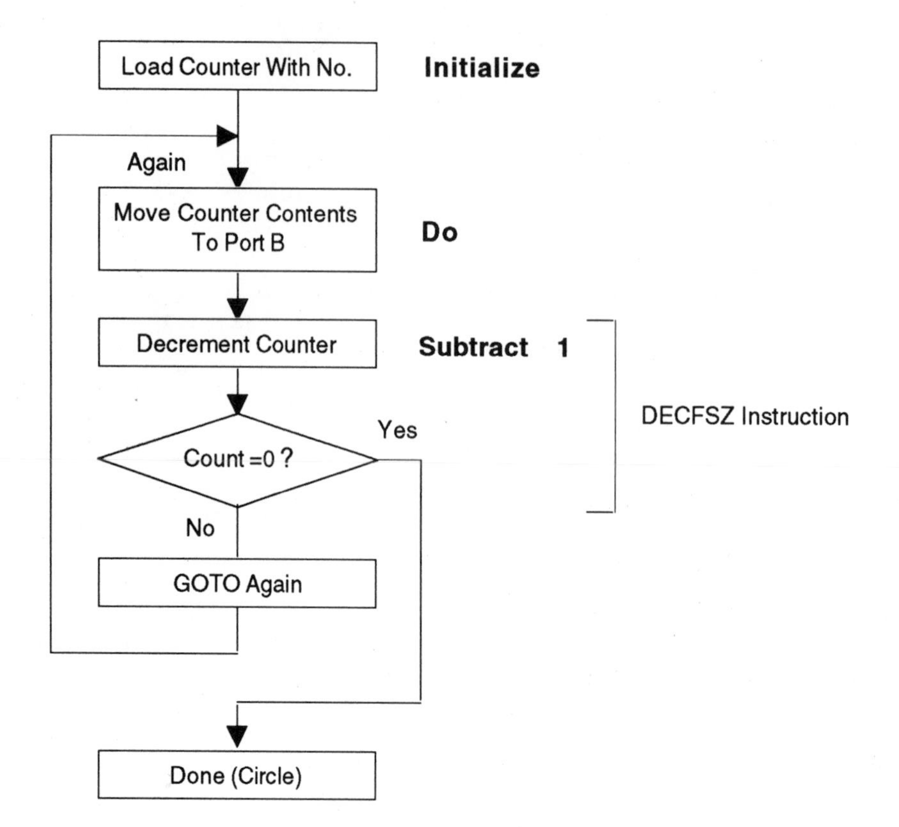


```
;=======PICT22.ASM==========================6/29/96==
;loop with counter demo
;       count down
;----------------------------------------------------
        list    p=16c84
        radix   hex
;----------------------------------------------------
;       destination designator equates
w       equ     0
f       equ     1
;----------------------------------------------------
;       cpu equates (memory map)
portb   equ     0x06
count   equ     0x0c
;----------------------------------------------------
        org     0x000
start   movlw   0x00    ;load w with 0x00
        tris    portb   ;copy w tristate, port B
;                           outputs
        movlw   0x05    ;load w with 0x05
        movwf   portb   ;display at port B
        movwf   count   ;W contents to counter
again   decfsz  count,f ;dec counter, result in count
        goto    again   ;not = 0x00
```

```
        movf    count,w ;get counter contents
        movwf   portb   ;counter contents to port B
circle  goto    circle  ;= 0x00, done
;
        end
;------------------------------------------------------
;at blast time, select:
;       memory unprotected
;       watchdog timer disabled (default is enabled)
;       standard crystal (using 4 MHz osc for test)
;       power-up timer on
;======================================================
```

Port B will indicate 0x00 in binary (00000000). Since nothing comes on, this demo is boring, but it illustrates the concept. This technique will be used in time delay loops later.

A key consideration in designing program loops using counters is what order to do things in. The three things going on are doing whatever the loop is designed to do, incrementing the counter and testing the counter. If these operations are done in the wrong order, the program will produce strange results.

A special situation arises in programs designed to transfer data or access tables. If operations are not done in the proper order, the program may fail to move or test the last byte. This all centers around whether or not the 0th operation or location counts as 0 or 1. The correct sequence is:

```
Do                  Use bit test, skip if
Comparison/test     Will have completed
Increment              no. of operations =
Goto                   counter contents).
Done
```

LOOP UNTIL

Another use for a loop is a loop-until situation.

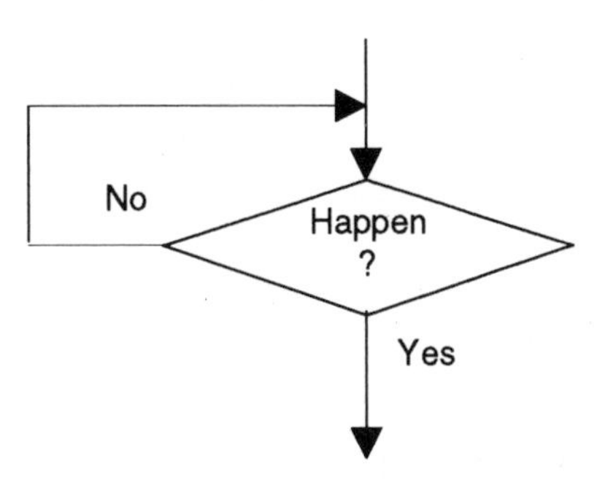

An example follows in the section on bit manipulation. The program goes into a loop until a switch is opened.

COMPARISONS

The contents of a file register can be compared with the contents of the W register to determine their relative magnitudes. This is done this by subtracting the contents of W from the contents of the selected register.

```
         movlw      0xXX       ;load W with literal
         subwf      selected   ;subtract contents of W from
;                                 contents of selected file reg,
;                                 result in file reg
         btfsc/s    status,2   ;test Z-flag, skip next instruction
;                                 if result 0/1
;        goto
```

Result:

```
     Carry flag is set if      W ≤ f
     Carry flag is clear if    W > f

     Zero flag is set if       W = f
     Zero flag is clear if     W ≠ f
```

A branch may be taken or not depending on the results of the comparison.

Test For	Use Subtraction Followed By	Flag Tested
W = f	BTFSS	Z Set
W ≠ f	BTFSC	Z Clear
W ≤ f	BTFSS	C Set
W > f	BTFSC	C Clear

Flags: Set = 1, Clear = 0

Follow the subtraction with "bit test, skip if" to send the microcontroller to whatever code is appropriate depending on the results of the comparison.

Here is a simple program to demonstrate a method of testing the comparison procedure. LED's are assumed to be connected to port "B" for use as an indicator.

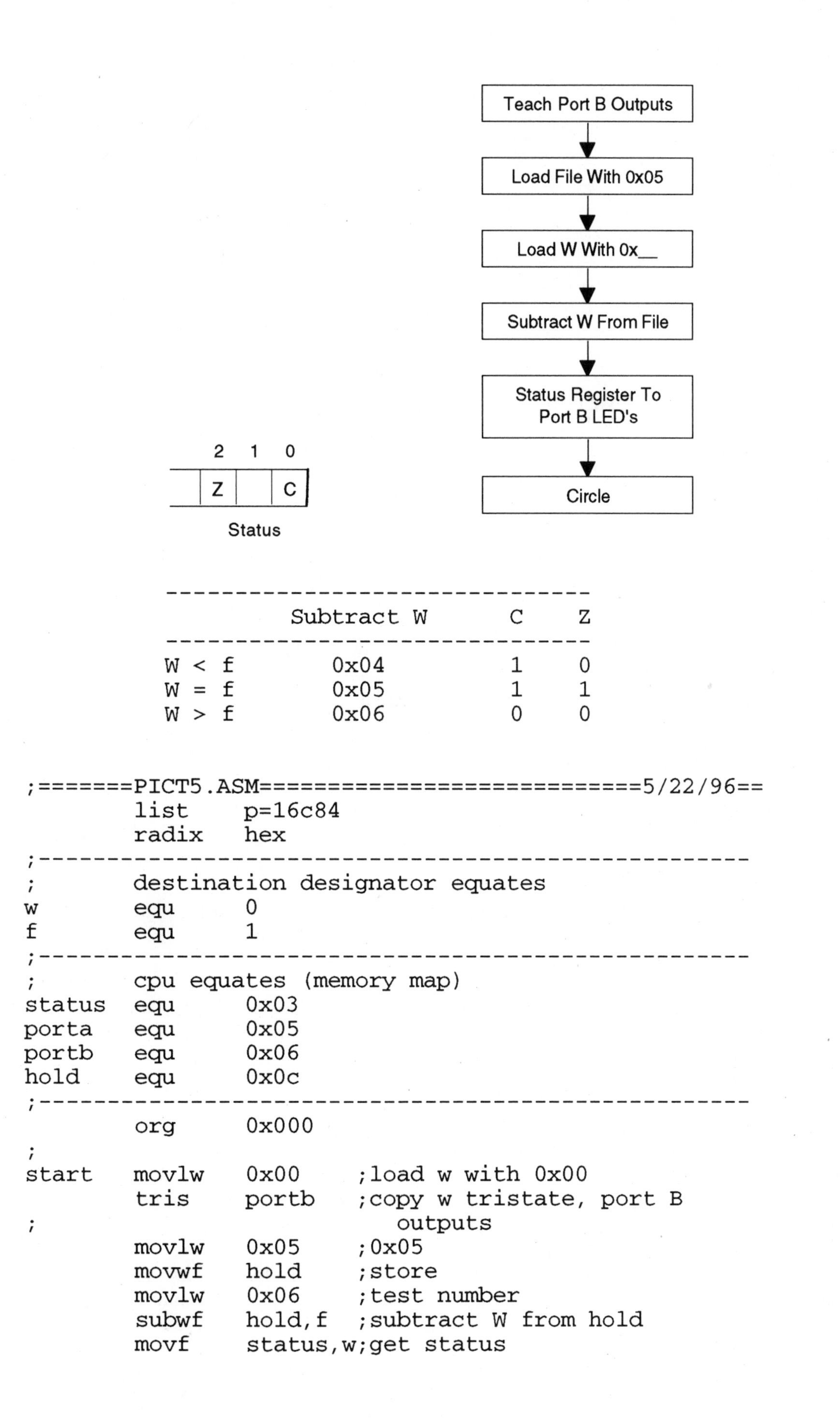


```
--------------------------------
           Subtract W      C    Z
--------------------------------
W < f        0x04          1    0
W = f        0x05          1    1
W > f        0x06          0    0
```

```
;=======PICT5.ASM===========================5/22/96==
        list    p=16c84
        radix   hex
;----------------------------------------------------
;       destination designator equates
w       equ     0
f       equ     1
;----------------------------------------------------
;       cpu equates (memory map)
status  equ     0x03
porta   equ     0x05
portb   equ     0x06
hold    equ     0x0c
;----------------------------------------------------
        org     0x000
;
start   movlw   0x00     ;load w with 0x00
        tris    portb    ;copy w tristate, port B
;                           outputs
        movlw   0x05     ;0x05
        movwf   hold     ;store
        movlw   0x06     ;test number
        subwf   hold,f   ;subtract W from hold
        movf    status,w;get status
```

```
        movwf   portb   ;display status via LED's
circle  goto    circle  ;done
;
        end
;-----------------------------------------------------
;at blast time, select:
;       memory unprotected
;       watchdog timer disabled (default is enabled)
;       standard crystal (using 4 MHz osc for test)
;       power-up timer on
;=====================================================
```

Note that subtraction followed by a check to see if the carry flag is set tests for W "less than"or"equal to" f. To test for W "less than" f exclusively, use the desired test number (N) minus one in W. W = M = N-1.

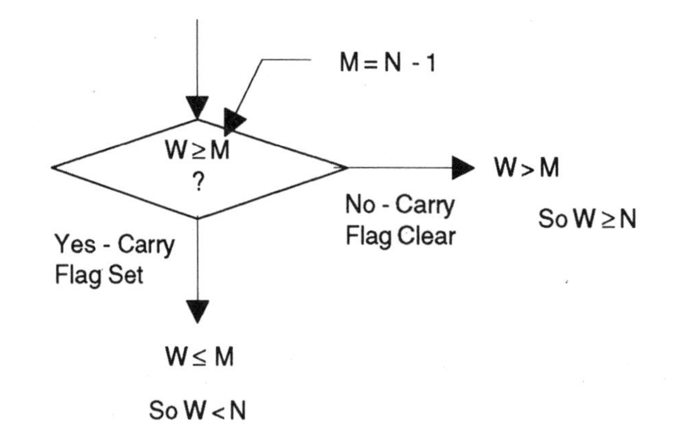

To show how this works, let's assume 0x05 is the target and we want to test for smaller numbers Use the test program above.

```
-----------------------------------------
W register contents    W=M=(N-1)    C Flag
-----------------------------------------
      W = 0x03        0x03 - 0x04    Set
      W = 0x04        0x04 - 0x04    Set
      W = 0x05        0x05 - 0x04    Clear
```

To review - test stuff by comparing an unknown with a known standard and branching to a set of instructions located elsewhere if the test is affirmative.

Another way to test for two bytes being equal is to use the XORWL instruction. The byte to be tested is XOR'd with a test byte. If they match, the Z-flag will be set (see bit manipulation using logic instructions section).

A method of testing a byte in a file register to see if it is zero is to use the MOVF instruction and place the result in the register it came from. The Z-flag is effected, so if it is set after the byte is replaced, the register contains "0".

```
movf        reg,f       ;move file, replace
btfss       status,2    ;skip next if Z set
goto        elsewhere
continue
```

BIT MANIPULATION USING BIT MANIPULATION INSTRUCTIONS

The PIC16C84 has bit manipulation instructions which can be used to set or clear an individual bit or to test an individual bit. These instructions operate on the file registers.

Bit Set/Clear

Bit set (BSF) and bit clear (BCF) instructions operate on a selected bit in a selected register.

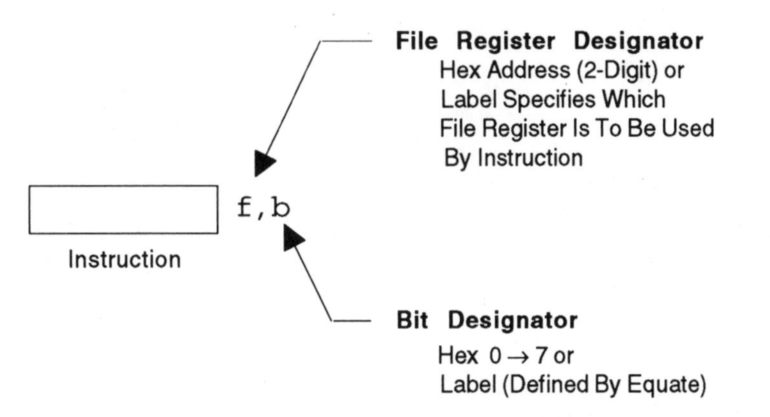

As an example, bit 3 of port A can be set (made logic 1 or HI) by:

```
bsf         porta,3     ;set port A, bit 3
```

Bit Testing

A bit in a file register may be tested using the BTFSC or BTFSS instruction.

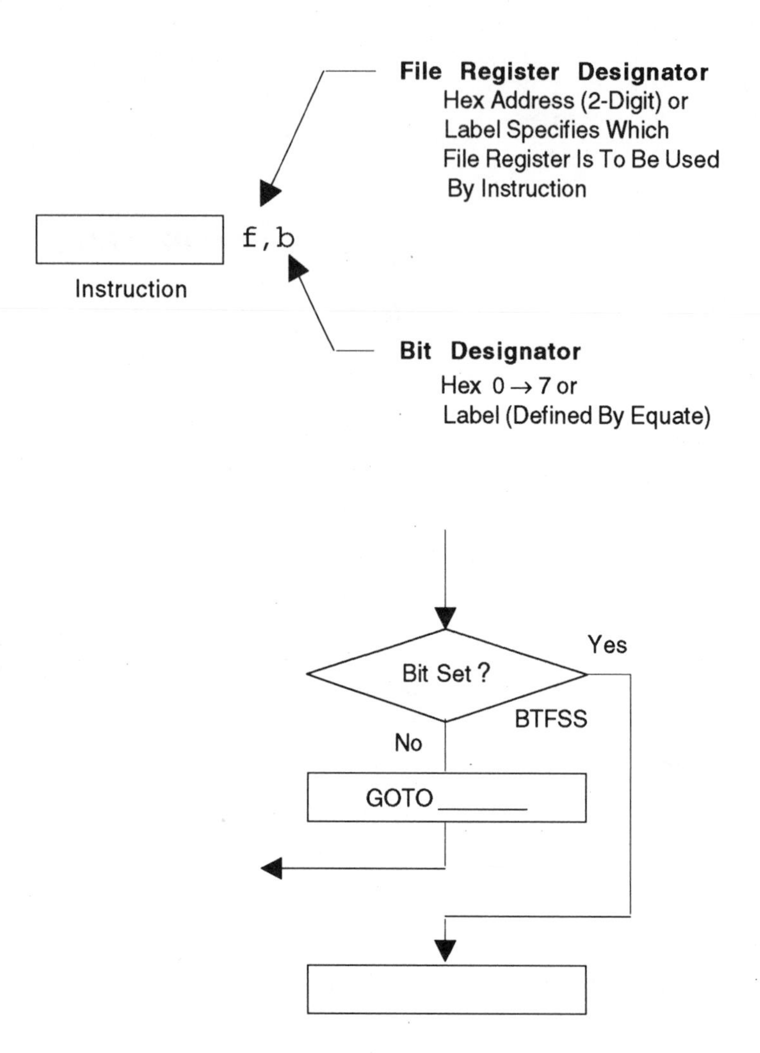

By way of example, we want to test bit 0 in the file register labeled "indicator" to see if it is clear (0).

```
          btfsc     indicator,0         ;test indicator bit 0
;                                           skip next instruction
;                                           if clear
          goto      elsewhere           ;jump if set
          continue
```

If the bit tested is clear, the next instruction is skipped. So the GOTO is a branch which will be taken if bit 0 is set. Otherwise, code execution continues in sequence.

BIT MANIPULATION USING LOGIC INSTRUCTIONS

Logical operations and shifting bytes sideways may be used to change or test specific bits in a file register or in the W register. These methods are useful when two or more bits in a byte must be changed or tested in one instruction cycle.

Change Specific Bit To "1"

```
OR     with an 8-bit binary number which is all zeros except
          the bit to be changed to a "1"
OR     with "0" leaves bit unchanged
OR     with "1" results in "1"
```

Change Specific Bit To "0"

```
AND    with an 8-bit binary number which is all ones except
          the bit to be changed to a "0"
AND    with "1" leaves bit unchanged
AND    with "0" results in "0" (called masking)
```

Change Specific Bit To It's Complement

```
0 to 1 & 1 to 0

XOR    with an 8-bit binary number which is all zero's except
          the bit to be changed to it's complement (Exclusive
          OR).
XOR    with "1" changes bit to complement
XOR    with "0" results in no change
```

Comparison - Test For Specific Byte

```
XOR    with byte = to the one you are looking for.  If Z-flag is
          set, they match.  This is another comparison method to be
          used when looking for identical bytes.
```

Test For "0"

```
OR     with "0", then test the Z-flag.  If the Z-flag is set, the
          bytes match and the byte tested is "0".
```

We can try out the logic bit manipulation instructions by playing with bit 5 (arbitrary choice) in the W register.

```
        Make it a "1" with  ORA 0010 0000

          movlw     b'00000000'
          iorwl     b'00100000'      ;make bit 5 a 1
          movwf     portb            ;display at port b
circle    goto      circle           ;circle


        Make it a "0" with  AND 1101 1111

          movlw     b'00100000'
          andlw     b'11011111'      ;make bit 5 a 0
          movwf     portb            ;display at port b
circle    goto      circle           ;circle


        Change it with      XOR 0010 0000    (compliment or
                                              invert)

          movlw     b'00000000'
          xorlw     b'00100000'      ;compliment bit 5
          movwf     portb            ;display at port b
circle    goto      circle           ;circle


        Test for equal bytes

          movlw     b'01010101'
          xorlw     b'01010101'
          movwf     portb            ;display at port b
circle    goto      circle           ;circle


        Test for "0"

          movlw     b'00000000'      ;test byte
          iorlw     b'00000000'      ;OR with "0"
          movf      status,w         ;get status
          movwf     portb            ;display   ;status
circle    goto      circle           ;circle
```

USING BIT MANIPULATION

Next we will try a program which uses bit manipulation. The objective of the program is to look at an input port line and when it has a 1 on it, output a 1 on an output port line (if sense X, then turn on Y). Bit 3 of port A is arbitrarily chosen as the input line to be sensed and line bit 2 of the port B is arbitrarily chosen as the line to be turned on. The input line switch is closed before the program is run. The output port line is set to 0 at the beginning of the program so the LED will be off when the program is initiated.

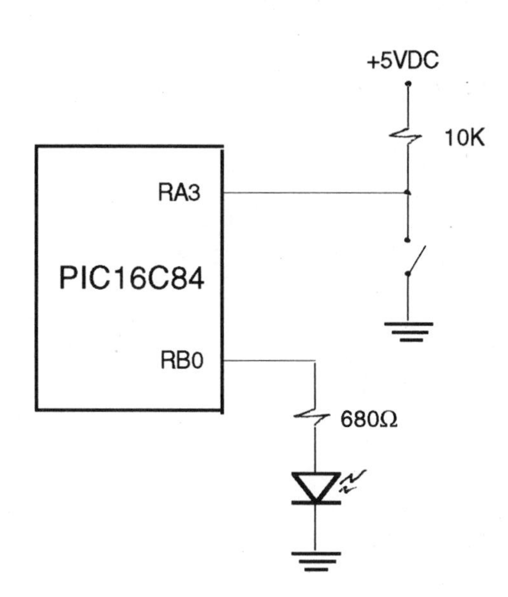

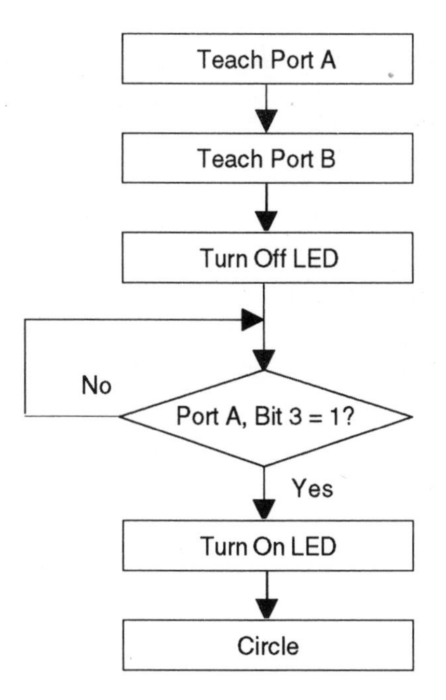

```
;=======PICT10.ASM==========================6/25/96==
;bit manipulation demo
;----------------------------------------------------
        list    p=16c84
        radix   hex
;----------------------------------------------------
;       destination designator equates
w       equ     0
f       equ     1
;----------------------------------------------------
;       cpu equates (memory map)
porta   equ     0x05
portb   equ     0x06
;----------------------------------------------------
        org     0x000
;
start   movlw   b'11111111'  ;load W
        tris    porta   ;teach port A
        movlw   b'11111011'  ;load W
        tris    portb   ;teach port B
        bcf     portb,2 ;turn off LED
switch  btfss   porta,3 ;bit 3 HI?
        goto    switch  ;no
        bsf     portb,2 ;yes, turn on LED
circle  goto    circle  ;done
;
        end
;----------------------------------------------------
;at blast time, select:
;       memory unprotected
;       watchdog timer disabled (default is enabled)
;       standard crystal (using 4 MHz osc for test)
;       power-up timer on
;====================================================
```

The next example program illustrates an application for bit manipulation in event counting. Events represented by a series of switch closures can be counted. The example program will count the number of times the bit 3 switch used in the previous example is opened allowing the port A bit 3 input line to be pulled up to logic 1. The count is displayed in binary on LED's connected to port B.

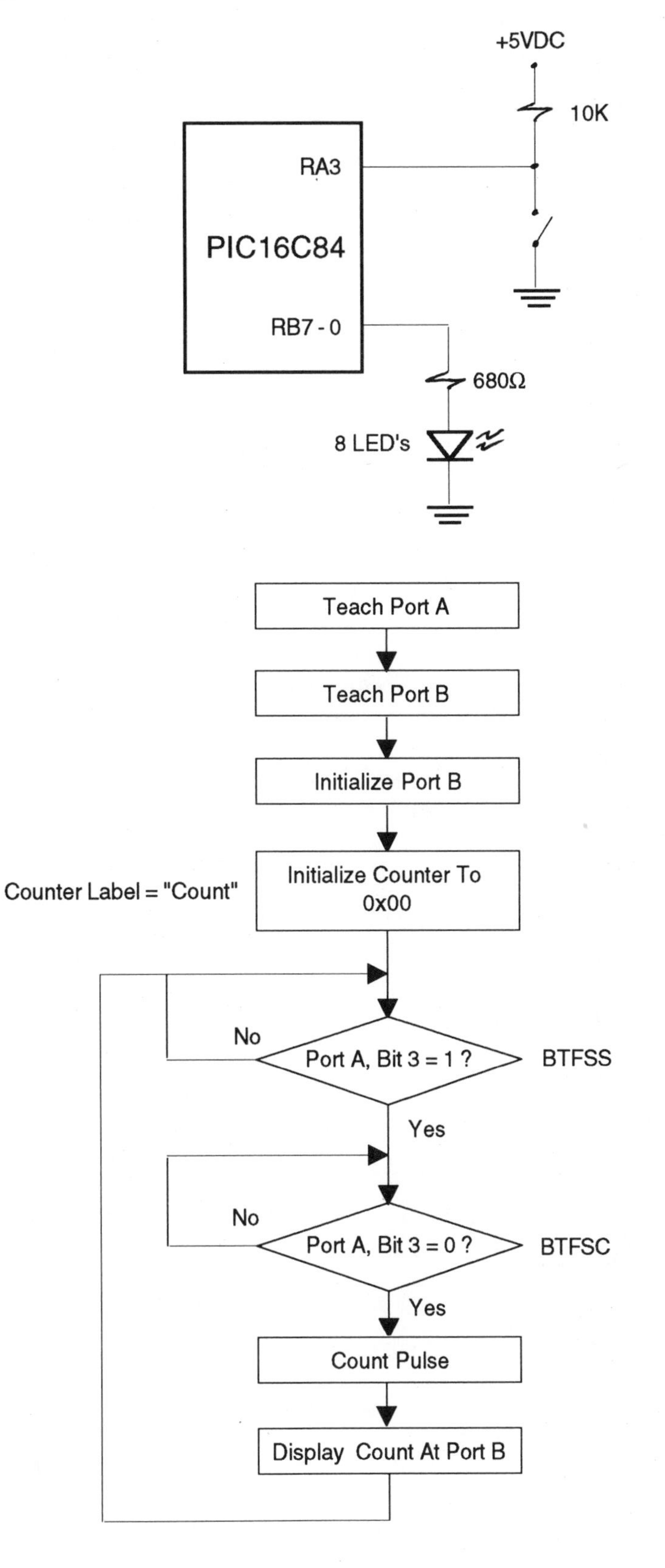

Start with switch closed, bit = 0.

Loop 'til bit = 1 which means a transition from 0 to 1 has occurred (leading edge of pulse detected).

Loop 'til bit = 0 which means a transition from 1 to 0 has occurred (trailing edge of pulse detected).

The pulse is counted when the trailing edge has been detected.

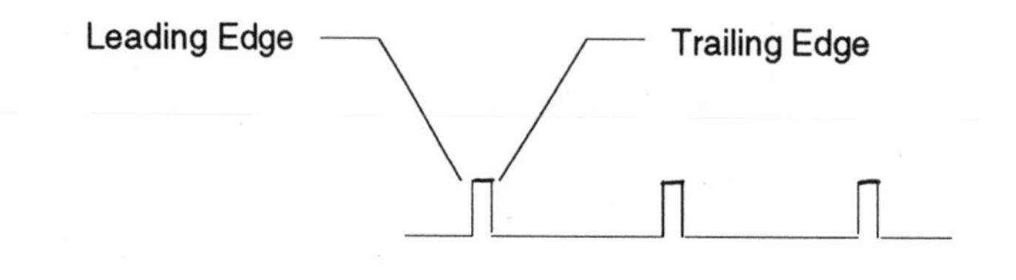


```
;=======PICT11.ASM============================6/25/96==
;event counting demo
;---------------------------------------------------
        list    p=16c84
        radix   hex
;----------------------------------------------------
;       destination designator equates
w       equ     0
f       equ     1
;----------------------------------------------------
;       cpu equates (memory map)
porta   equ     0x05
portb   equ     0x06
count   equ     0x0c
;----------------------------------------------------
        org     0x000
;
start   movlw   0xff
        tris    porta   ;teach port A inputs
        movlw   0x00
        tris    portb   ;teach port B outputs
        clrf    portb   ;port B lines LO
        clrf    count   ;initialize counter
hi?     btfss   porta,3 ;port A, bit 3 HI?
        goto    hi?     ;no
lo?     btfsc   porta,3 ;no
        goto    lo?     ;no
        incf    count,f ;yes
        movf    count,w ;counter contents to W
        movwf   portb   ;display counter contents
;                           at port B
        goto    hi?     ;again
```

SEQUENCING

The RLF (rotate file register contents left one bit) instruction can be used to shift data memory contents left one bit position at a time. Each time the RLF instruction is executed, the byte is shifted left one position, the left or most significant bit is moved to the carry flag, and the carry flag contents is shifted to the right or least significant bit.

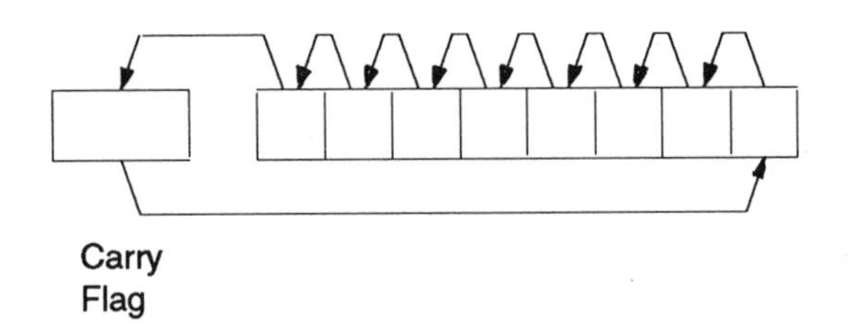

The RRF instruction does the same thing, but in the opposite direction.

Sometimes it is necessary to circulate a bit pattern. This might be a requirement in the operation of production machinery or in driving a light display.

The RLF instruction maintains the initial bit pattern by cycling the bits out through the carry flag and back into the least significant bit position.

Since the carry flag contents get recycled when the RLF instruction is used, we must start our program by putting the carry flag in a known state ("1" or "0"). Changing the carry flag contents is done using a bit set or bit clear instruction to change bit 0 of the status register (carry flag).

```
status      equ       0x03        ;status word register
c           equ       0           ;bit 0 is carry flag
file        equ       0x0c        ;file register w/ bit
                                      pattern
;
;
            bcf       status,c    ;clear carry flag
            rlf       file,f      ;rotate bits left,
                                      result in file
;
```

To show how this works, let's write a simple program which cycles a bit pattern through the eight LED's at port B (output). We will need a time delay to slow things down enough so we can see what's happening, so we will use the "pause" time delay subroutine (operation explained in the Timing and Counting chapter). We will start off as shown and rotate left (the RRF instruction could be used to rotate right).

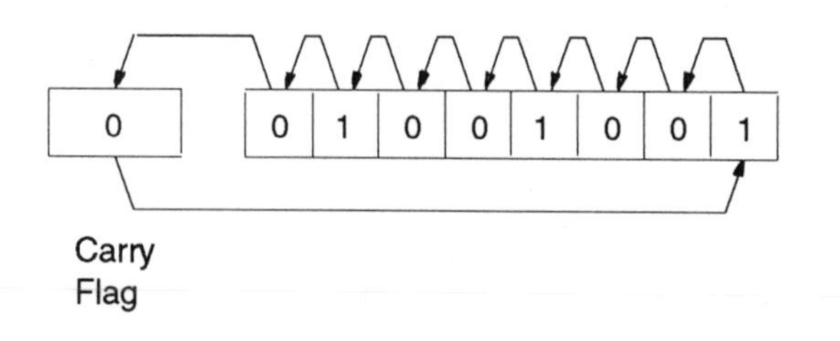


```
;=======PICT13.ASM============================6/27/96==
;sequencing demo
;       uses time delay "pause"
;----------------------------------------------------
        list    p=16c84
        radix   hex
;----------------------------------------------------
;       destination designator equates
w       equ     0
f       equ     1
;----------------------------------------------------
;       cpu equates (memory map)
status  equ     0x03
portb   equ     0x06
shift   equ     0x0c
ncount  equ     0x0d
mcount  equ     0x0e
;----------------------------------------------------
;       bit equates
c       equ     0
;----------------------------------------------------
        org     0x000
;
start   movlw   0x00    ;teach port B outputs
        tris    portb
        clrf    portb   ;all lines low
        bcf     status,c  ;clear carry flag
        movlw   b'01001001'  ;initial bit pattern
        movwf   shift   ;store in shift
get_bit movf    shift,w ;get bit pattern
        movwf   portb   ;display
        call    pause   ;delay via subroutine
        rlf     shift,f ;rotate bits
        goto    get_bit ;again
;
```

```
pause   movlw   0xff     ;M
        movwf   mcount   ;to M counter
loadn   movlw   0xff     ;N
        movwf   ncount   ;to N counter
decn    decfsz  ncount,f   ;decrement N
        goto    decn     ;again
        decfsz  mcount,f   ;decrement M
        goto    loadn    ;again
        return           ;done
;
        end
;------------------------------------------------------
;at blast time, select:
;       memory unprotected
;       watchdog timer disabled (default is enabled)
;       standard crystal (using 4 MHz osc for test)
;       power-up timer on
;======================================================
```

If 8 bits plus the carry flag is not enough, how about 16? A double-wide RLF sequence might look like this:

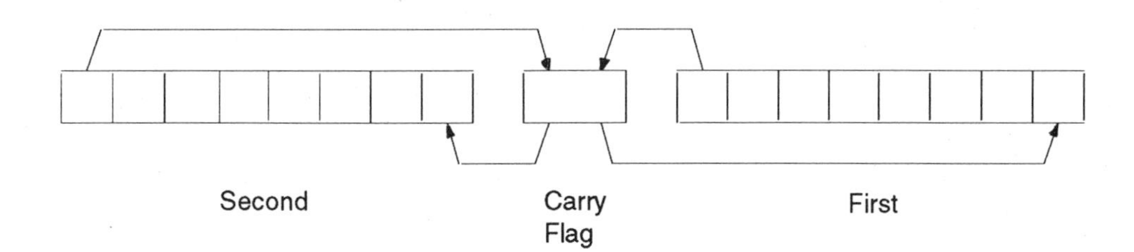

You will need two memory locations for this one and the carry flag must be cleared or set when the program starts.

The first RLF operating on the memory location shown on the right side of the drawing stuffs the most significant bit into the carry flag and the carry flag contents into the least significant bit of the right side memory location.

The second RLF instruction operating on the memory location shown on the left pulls the carry flag bit into the least significant bit and pushes the most significant bit into the carry flag.

This sequence can be repeated indefinitely. Assuming a larger microcontroller, the contents of the two memory locations would be sent to two parallel ports after each pair of RLF instructions have been executed to update the status of the sixteen output lines.

SUBROUTINES

The next example will illustrate the use of time delays and subroutines. The program counts in binary and displays the count at eight LED's via port B. If the counting were done at full speed, the LED's would appear to be all on all of the time. Instead, the time delay subroutine "pause" will be used to provide a time delay between counts so that each count will be displayed for about one second. It delays for 0.2 second (200 milliseconds). A call subroutine instruction (CALL) is used to call the "pause" subroutine. The routine is used five times in succession to obtain a 1 second delay. The last instruction in the time delay subroutine is return for subroutine (RETURN).

The details of the "pause" time delay subroutine will be explained in the next section (time delay loop).

A file register called "count" will be used as a counter.

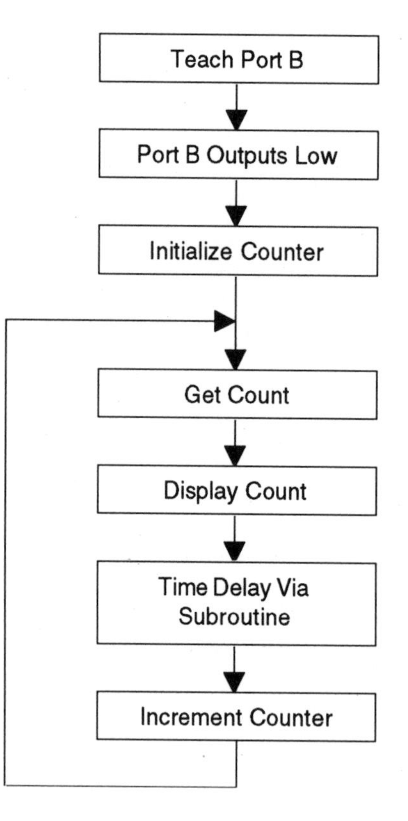

```
;=======PICT12.ASM===========================6/27/96==
;subroutine demo
;       uses time delay "pause"
;---------------------------------------------------
        list    p=16c84
        radix   hex
;---------------------------------------------------
;       destination designator equates
w       equ     0
f       equ     1
;---------------------------------------------------
;       cpu equates (memory map)
portb   equ     0x06
count   equ     0x0c
ncount  equ     0x0d
mcount  equ     0x0e
;---------------------------------------------------
        org     0x000
;
start   movlw   0x00     ;teach port B outputs
        tris    portb
        clrf    portb    ;all lines low
        clrf    count    ;initialize counter
get_cnt movf    count,w  ;get count
        movwf   portb    ;display count
        call    pause    ;delay via subroutine
        call    pause
        call    pause
        call    pause
        call    pause
        incf    count,f  ;increment counter
        goto    get_cnt  ;again
;
pause   movlw   0xff     ;M
        movwf   mcount   ;to M counter
loadn   movlw   0xff     ;N
        movwf   ncount   ;to N counter
decn    decfsz  ncount,f   ;decrement N
        goto    decn     ;again
        decfsz  mcount,f   ;decrement M
        goto    loadn    ;again
        return           ;done
;
        end
;---------------------------------------------------
;at blast time, select:
;       memory unprotected
;       watchdog timer disabled (default is enabled)
;       standard crystal (using 4 MHz osc for test)
;       power-up timer on
;===================================================
```

Note that this program loops until the reset switch is pressed.

The "count" register is an 8-bit counter which has $2^8 = 256$ possible combinations of 0's and 1's.

Subroutines are very useful. You may think of them as program modules which do some specific task. Once you have written or otherwise obtained a useful subroutine, it is yours to incorporate into future programs. This saves a lot of work. Also writing and debugging one subroutine at a time is much easier than trying to write and debug a large program all at one go.

Your main program can call one subroutine after another to do a large job. Also one subroutine can call another. As long as each subroutine ends with an RETURN or RETLW instruction, the sequence of operations will work its way back to the main program.

The microcontroller uses a hardware stack as a place to keep track of program addresses as it moves into and back out of subroutines. The operation of the stack for subroutine return addresses is automatic.

Remember that a subroutine must end with a RETURN or RETLW instruction.

The PIC16C84 stack is eight deep, meaning that the main program can call one subroutine which calls a second subroutine, etc. The limit is eight deep.

TIME DELAY LOOP

Sometimes it is necessary to wait a specified period of time for something to happen. This might be waiting for switch contacts to stop bouncing or holding an EPROM control line high for a specified time as part of the programming procedure. Time intervals can be measured using hardware (covered later) or by software using a time delay loop. A time delay loop counts clock cycles until a time interval has elapsed. The accuracy of the interval depends on the accuracy of the PIC16C84 system clock. For accurate delays, a crystal-controlled clock oscillator is recommended. If a 4 megahertz crystal oscillator is used, the oscillator input to the PIC16C84 is divided by 4 inside the PIC16C84. So the internal clock frequency is 1 megahertz.

Each instruction executed by the PIC16C84 takes a set number of clock cycles. We will use a file register as a counter. The number of internal clock cycles required for the instructions we will be using is shown in the table:

Instruction	Clock Cycles
MOVLW	1
MOVFW	1
DECFSZ	1, 2 if result is 0
GOTO	2
CALL	2
RETURN	2

PIC16/17 instructions execute in one instruction cycle except:

- GOTO
- CALL
- Instructions which write to the program counter

These exceptions require two instruction cycles to execute.

A simple time delay loop using a file register follows:

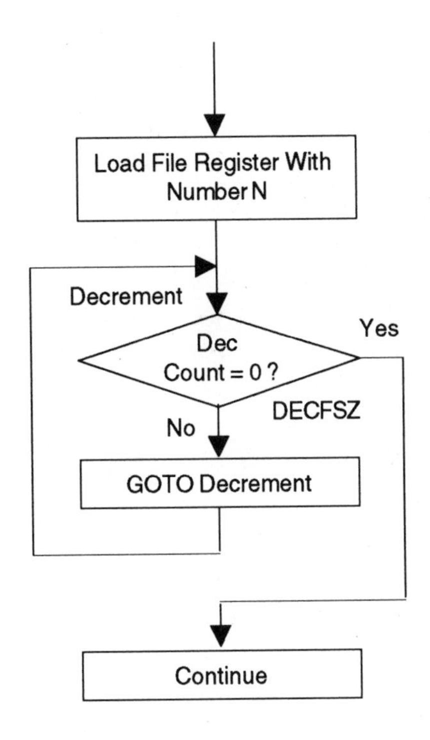


```
--------
Internal
Clock
Cycles
--------
  1         MOVLW          Load counter
  1         MOVWF
  1xN       DECFSZ  *      Decrement counter
  2xN       GOTO           Done ?

            * 2 instruction cycles required last
              time through (count = 0)
```

The number of clock cycles required to run this delay program depends on the number loaded into the counter at the beginning of the program (N).

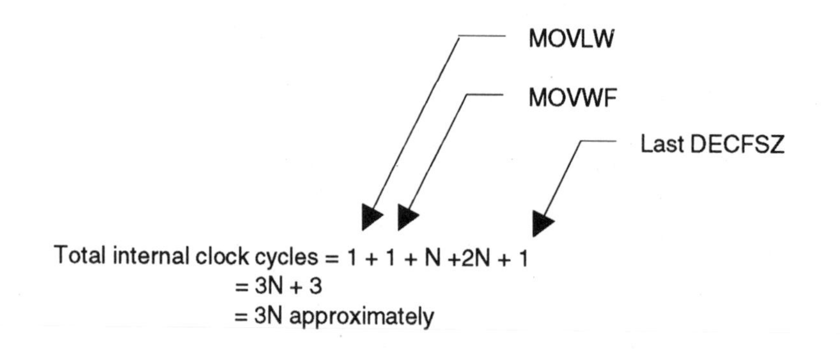

If the PIC16C84 internal clock frequency is 1 megahertz, each clock cycle is 1 microsecond. If the hexadecimal equivalent of decimal 50 were loaded into the counter, the program would produce a delay of

```
3(50)  +2 = 152 internal clock cycles
          = 152 microseconds
```

If the program above were used as a subroutine, 2 additional cycles would be required for the CALL instruction and 2 more for the RETURN instruction for a total of 4 additional cycles.

The maximum delay possible with this program is achieved by loading 0x00, strangely enough. The decrement instruction will go through to 0xFF and so on. The maximum delay is

```
3(256) +2 = 770 cycles
          = 770 microseconds (assuming 4 MHz crystal-
                controlled clock oscillator)
```

It is best not to agonize over the math. Calculate an approximate number and run the program. Measure the result and change the number if necessary, then try again.

Using a program with a time delay subroutine to turn output port B on and off will serve as a demonstration. Connect an oscilloscope to port B to measure the width of the pulses.

```
;=======PICT8.ASM=============================6/25/96==
;time delay demo
;       file register counter - count to 10
;       delay = 10+  microseconds
;----------------------------------------------------
        list    p=16c84
        radix   hex
;----------------------------------------------------
;       destination designator equates
w       equ     0
f       equ     1
;----------------------------------------------------
;       cpu equates (memory map)
portb   equ     0x06
count   equ     0x0c
;----------------------------------------------------
        org     0X000
;
start   movlw   0x00     ;load w with 0x00
        tris    portb    ;copy w tristate, port B
;                           outputs
        clrf    portb    ;all lines low
go      bsf     portb,0  ;turn on LED
        call    delay    ;delay via sub
        bcf     portb,0  ;turn off LED
        call    delay    ;delay via sub
        goto    go       ;repeat
;
delay   movlw   0x0a     ;decimal 10
        movwf   count    ;load counter
repeat  decfsz  count,f  ;decrement counter
        goto    repeat   ;not 0
        return           ;counter 0, end delay
;
        end
;----------------------------------------------------
;at blast time, select:
;       memory unprotected
;       watchdog timer disabled (default is enabled)
;       standard crystal (using 4 MHz osc for test)
;       power-up timer on
;====================================================
```

```
call      2
movlw     1
movwf     1
decfsz    1xN
goto      2xN
decfsz    1      (last)
return    2
          7+3N clock cycles or approx. 3N clock cycles
```

Run the program. Try various hex numbers and measure the time delays. 0x1E gives a delay of approximately 100 microseconds. Notice that the LO interval is 3 microseconds longer than the HI. This is because the GOTO instruction requires 2 microseconds and the two moves take 2 microseconds (turn on) vs. 1 microsecond for the CLRF instruction (turn off).

Sample calculation:

```
                   2     Turn on
     Goes HI
                 100     Delay 7 + 3(31) = 100
                   1     Turn off
     Goes LO
                 100     Delay
                   2     GOTO

              101 on, 104 off
```

A delay of 200 milliseconds (called "pause") is used for examples in this book. It is generated using nested loops (loop within a loop). Two counters are required.

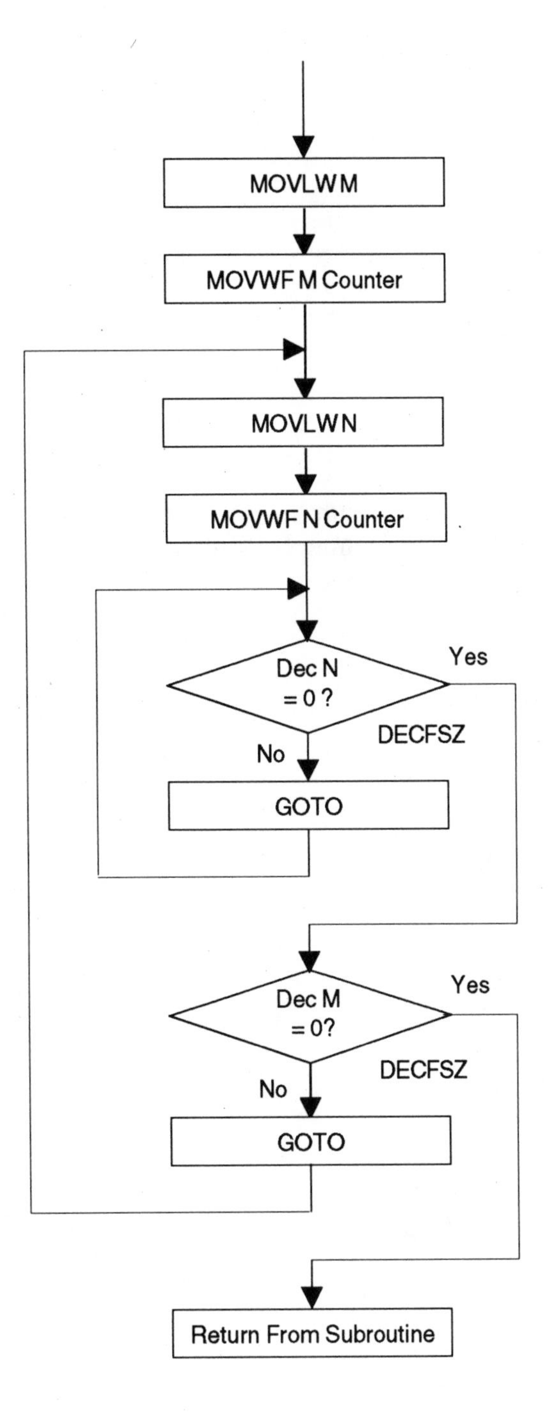

```
--------
Internal
Clock
Cycles
--------
 2          CALL
 1          MOVLW M       Load M
 1          MOVWF         In M counter
 1xM        MOVLW N       Load N
 1xM        MOVWF         IN N counter
 1xNxM      DECFSZ        Decrement N counter
 2xNxM      GOTO          N = 0 ?
 1xM        DECFSZ        Decrement M counter
 2xM        GOTO          M = 0 ?
 2          RETURN
```

Total cycles = approximately 3MN.

The maximum possible delay obtainable with this program is 3(256)(256) = 196,608 microseconds = 197 milliseconds (approximately) with a 1 megahertz internal clock.

Here is the loop-within-a-loop time delay subroutine "pause" used earlier.

```
ncount      equ         0x0d
mcount      equ         0x0e
;
pause       movlw       0x__        ;M
            movwf       mcount      ;to M counter
loadn       movlw       0x__        ;N
            movwf       ncount      ;to N counter
decn        decfsz      ncount,f    ;decrement N
            goto        decn        ;again
            decfsz      mcount,f    ;decrement M
            goto        loadn       ;again
            return                  ;done
```

This program uses two file registers as counters. As an exercise, let's assume it is necessary to generate a 12 millisecond delay.

Time (millisec)	Hex Number N = M
12	0x__

Sample calculation:

Desired delay 12 millisec

M = N = Y

$3Y^2$ = 12000 microsec = 12 millisec

$$Y^2 = \frac{12000}{3} = 4000 \quad Y = 4000 = 63.2 \text{ decimal} = 0x3E$$

Testing a program to count in binary with the count displayed at the LED output port will give you another opportunity to try a time delay subroutine.

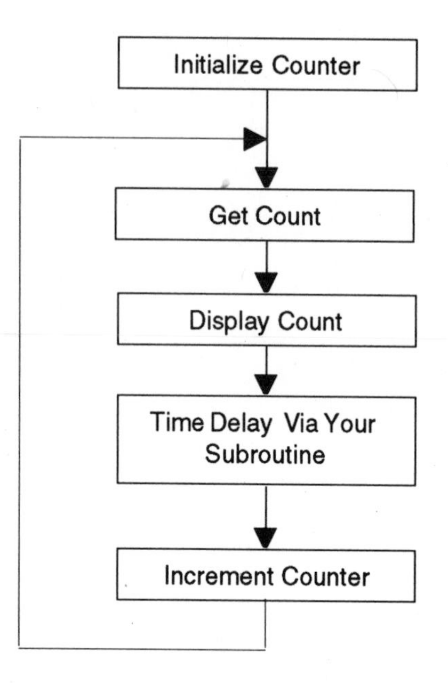


```
count       equ        0x0c
;
            movlw      0x00        ;teach port B outputs
            tris       portb
            clrf       portb       ;all lines low
            clrf       count       ;initialize counter
get_cnt     movf       count,w     ;get count
            movwf      portb       ;display count
            call       your_sub    ;your delay subroutine
            incf       cnt,f       ;increment counter
            goto       get_cnt     ;again
```

The delay should be long enough so you can see each count displayed on the LED's but not so long that you get bored waiting for the count to change.

LOOKUP TABLES

A lookup table may be used to convert one code to another. Let's say we want to convert hexadecimal characters to 7-segment signals to drive a display. We will assume that the hex character is in the least significant nybble of the byte.

HEX	7-SEGMENT CODE	DPgfe dcba
0x00	0x3F	0011 1111
01	06	0000 0110
02	5B	0101 1011
03	4F	0100 1111
04	66	0110 0110
05	6D	0110 1101
06	7D	0111 1101
07	07	0000 0111
08	7F	0111 1111
09	6F	0110 1111

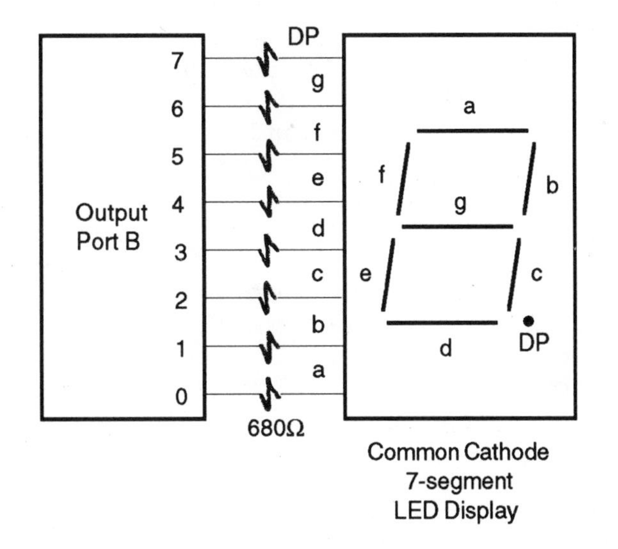

The proper 7-segment code may be pulled from the table using relative addressing. If the hex character is added to the base address of the table (program counter contents), the 7-segment code is stored at that address.

We will assume that the hex character to be converted is in the blank in the program line which begins with the label "char" and that the 7-segment equivalent is to be displayed.

```
;=======PICT14.ASM============================6/27/96==
;lookup table demo - 7-segment LED display
;-----------------------------------------------------
        list    p=16c84
        radix   hex
;-----------------------------------------------------
;       destination designator equates
w       equ     0
f       equ     1
;-----------------------------------------------------
;       cpu equates (memory map)
pc      equ     0x02    ;program counter
portb   equ     0x06
count   equ     0x0c    ;number
;-----------------------------------------------------
        org     0x000
;
start   movlw   0x00    ;load w with 0x00
        tris    portb   ;teach port B all outputs
        clrf    portb   ;all lines low
char    movlw   0x02    ;test number
        movwf   count   ;to file register count
;
        movf    count,w ;get count
        call    segmnt  ;call subroutine
        movwf   portb   ;display results
circle  goto    circle  ;done
;
segmnt  addwf   pc,f    ;add offset to program counter
        retlw   3f      ;0 seven segment
        retlw   06      ;1
        retlw   5b      ;2
        retlw   4f      ;3
        retlw   66      ;4
        retlw   6d      ;5
        retlw   7d      ;6
        retlw   07      ;7
        retlw   7f      ;8
        retlw   6f      ;9
;
        end
;-----------------------------------------------------
;at blast time, select:
;       memory unprotected
;       watchdog timer disabled (default is enabled)
;       standard crystal (using 4 MHz osc for test)
;       power-up timer on
;=====================================================
```

The processor adds the hex character to the program counter where it will be used as the index or offset to find the 7-segment code which is then displayed.

Store a hex character with a 0x0 in front of it (ie 0x04, etc.) in the program line which begins with the label "char". Run the program. Examine the 7-segment display to see if the program worked.

The ADDWF instruction adds 8 bits to the lower 8 bits of the program counter and does not effect the high 5 bits. The high 5 bits can be dealt with once program memory paging is understood. For now, tables must be located entirely within the first 256 words of program memory.

SUBROUTINE LIBRARY

I would encourage you to start a subroutine library. This can include subroutines you write to do various tasks and subroutines you find in magazine articles, on the Internet news groups or wherever. The subs can be saved and stored as text files for future use. This saves reinventing them each time they are needed.

TRY YOUR OWN EXPERIMENTS

I also encourage you to try some ideas on your own. You can combine some of the techniques from the examples such as count 0 to 9 with time delay and display the count on a 7-segment display. Use your imagination!

THERE'S MORE ----- LOTS MORE!!!!

The programming information given here is intended as an introduction and as a basis for designing and building your own microcontroller-controlled systems. The information is complete enough for many applications. If you can't find information you need here, I encourage you to consult Microchip's data books and their Embedded Control Handbook. One of the fun things about working with microcontrollers is that there is always more to learn and more exotic things that you can do

INTERRUPTS

When an event occurs which demands the microcontroller's attention, an interrupt may be generated which will cause the microcontroller to drop what it is doing, take care of the task that needs to be performed, and go back to what it was doing.

When an interrupt occurs, the instruction currently being executed is completed. Then the PIC16C84 jumps to address 0x004 in program memory and executes the instruction stored there. This program is called an interrupt service routine. An interrupt service routine may cause (as required) the microcontroller to first take notes on the status of the program it was executing when the interrupt occurred so that it can find its place when it comes back. Then the interrupt service routine will handle the interrupt by doing whatever needs to be done. On completion, the routine will review its notes, set everything back to the way it was and take up the main program where it left off.

Interrupts are caused by events which must trigger a response from the microcontroller at the time they occur. For the PIC16C84, interrupts may come from one of four sources:

- External - outside the microcontroller via the RB0/INT pin.
- Timer/counter TMR0 overflow from 0xFF to 0x00.
- Port B logic level change on bits 7,6,5,4.
- EEPROM write complete (not covered in this book because bank switching is required).

Interrupts may be enabled or disabled (masked) at two levels, global (all interrupts regardless of source) or specific (enable/disable specific interrupt sources).

INTERRUPT CONTROL REGISTER

The interrupt control register (INTCON) at address 0x0B handles enable/disable and contains the global interrupt flag and the flags for each interrupt source.

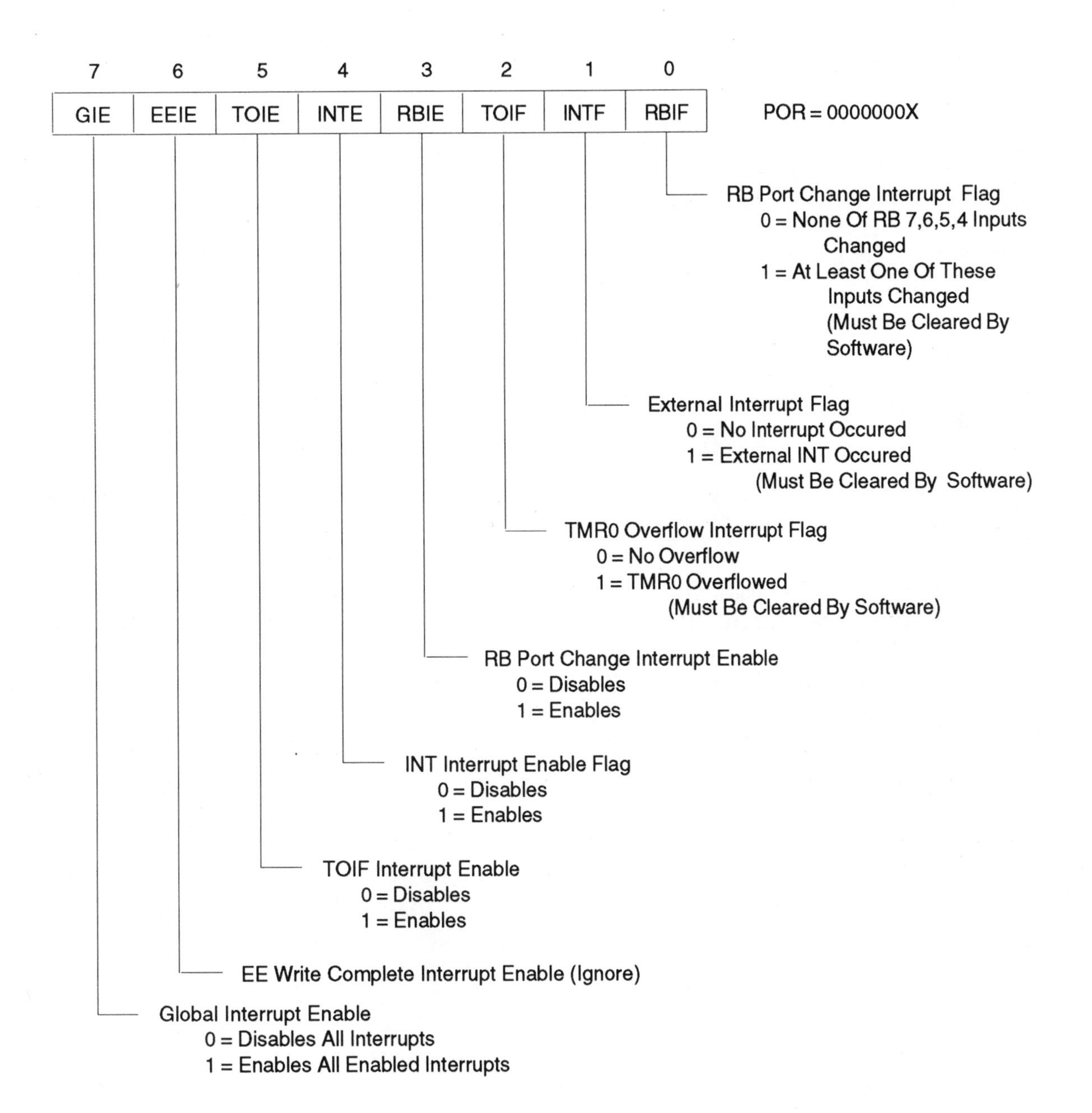

INTERRUPT SOURCES

External Interrupts

Using external interrupts requires both hardware and software. Hardware must be provided to sense an event or condition which requires an interrupt to bring about some action. A signal must be generated and fed to the microcontroller INT pin.

RBO/INT is a general purpose interrupt pin. It is edge-triggered. The INTEDG bit in the option register determines whether a rising or falling edge triggers an interrupt.

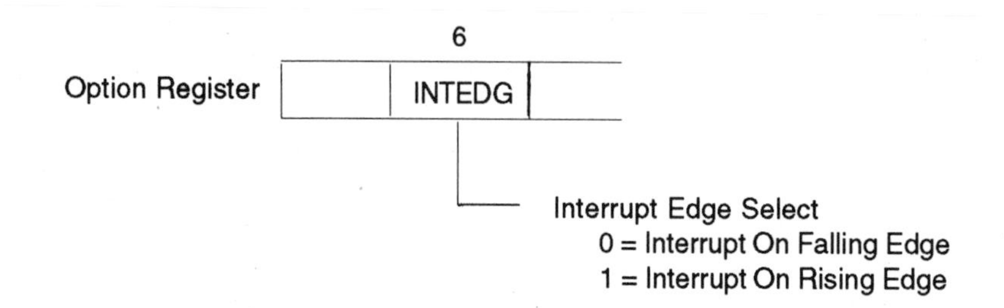

An INT interrupt can be disabled using the INT enable (INTE) flag in the INTCON register (bit 4) so that interrupts will be ignored until the INTE flag is set. INT interrupts can be serviced or ignored under software control.

If an INT interrupt occurs, the INTF flag (bit 1) in the INTCON register is set. The INTF flag must be cleared via software as part of the interrupt service routine before reenabling the interrupt or continuous interrupts will occur.

- Disable further interrupts (clear INTE flag)
- Service the interrupt
- Clear INTF interrupt flag
- Enable interrupts (set INTE flag)

Timer/Counter Interrupt

Timer/Counter TMR0 interrupts occur on overflow from 0xFF to 0x00. This is described in the Timing and Counting chapter.

Port B Interrupt On Change - Bits 7,6,5,4

An interrupt on logic level change on port B bits 7,6,5,4 sets INTCON bit 0 (RBIF). This interrupt is enabled/disabled via INTCON bit 3 (RBIE). Only port lines 7,6,5,4 configured as inputs are effected.

The pin's value in input mode is compared with the old value latched in the last reading of port B. The "mismatches" of the pins are OR'ed together to generate the RBIF interrupt (INTCON bit 0).
The interrupt may be cleared by:

- Disabling the interrupt by clearing the RBIE bit (INTCON bit 3).
- Read port B, then clear the RBIF bit. This ends the "mismatch" condition and allows RBIF to be cleared.

The interrupt on change feature is recommended for wakeup on key depression operation and operations where port B is only used for the interrupt on change feature. Polling of port B is not recommended while using the interrupt on change feature. Reading the port messes up the mismatch.

GLOBAL INTERRUPT ENABLE FLAG (GIE)

The occurance of an interrupt clears the global interrupt enable flag disabling further interrupts while the interrupt is being serviced. As the interrupt service routine is completed, execution of the return from interrupt (RETFIE) instruction sets the global interrupt flag enabling further interrupts.

SAVING STATUS DURING AN INTERRUPT (CONTEXT SAVING)

When an interrupt occurs, the program counter contents are saved on the stack automatically. The microcontroller needs to know where to resume code execution when servicing the interrupt is completed. Saving the status of other registers is the responsibility of the programmer and must be done to the extent the application requires as part of the interrupt service routine.

- Store W (must be saved first)
- Store status register
- Execute interrupt service
- Restore status register
- Restore W
- Return from interrupt (RETFIE)

```
save        movwf       temp_w      ;save w contents
            swapf       status,w    ;swap status, result in W
            movwf       temp_s      ;save status contents (swapped)
;
;Interrupt service ends with
;
            swapf       temp_s,w    ;reswap status, result in W
            movwf       status      ;restore status
            swapf       temp_w,f    ;swap, back in file
            swapf       temp_w,w    ;reswap, result in W
            retfie                  ;return with proper W contents
```

The SWAPF instruction will move data without effecting the status register (Z-flag). Of course the upper and lower nybbles get swapped in the process. The SWAPF instruction is needed to replace the status register contents without effecting it (can't use MOVF), so a swap must be made during the save to offset the swap which will occur on restore.

It took me quite a while to figure out why in the *@! SWAPF instructions were used in the examples of context-saving code I found. The MOVF instruction can't be used here because the Z-flag gets corrupted.

An interrupt service routine must end with a return from interrupt instruction (RETFIE). Execution of the RETFIE instruction causes the program counter to be loaded with the address saved on the stack when the interrupt occurred and execution resumes where it left off.

WHERE TO PUT THE INTERRUPT SERVICE ROUTINE

The interrupt vector built into the PIC16C84 is 0x004. When an interrupt occurs, the interrupt vector points to program memory address 0x004 where the first instruction of the interrupt service routine must be stored.

When interrupts are not used, the main program may start at address 0x000 and run right through 0x004 without conflict. If interrupts are used, other arrangements must be made by the programmer.

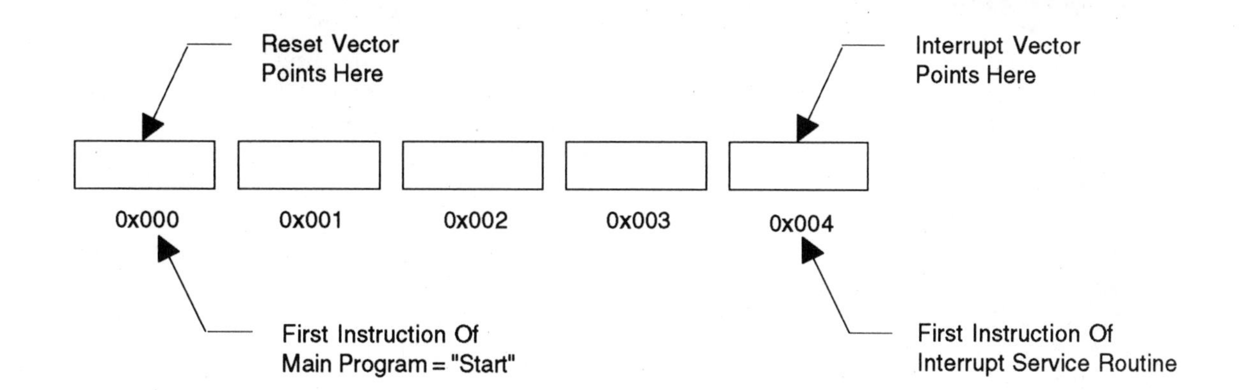

Probably the easiest thing to do is:

```
          org       0x000
          goto      start       ;real start of main program
;
          org       0x004
          goto      iservice    ;real start of interrupt service
;
start     your code
          ---------
          ---------
iservice  your routine
          ---------
```

Let the assembler locate the main program and interrupt service routine in program memory.

INTERRUPT LATENCY

When an interrupt occurs, there will be a delay (latency) before the interrupt service routine is executed. This delay will be 3 or 4 instruction cycles.

MULTIPLE EXTERNAL INTERRUPT SOURCES

If more than one interrupt is possible, the hardware will generate signals to interrupt the microcontroller and also tell it which function caused the interrupt. A different software subroutine may be necessary for each interrupt source.

A microcontroller such as the PIC16C84 may have only one interrupt line available for general system use. If more than one interrupt source is required, the microcontroller can poll an input port line associated with each interrupt source. Software checks each possible source to see which one needs service. Priorities can be assigned by selecting the order in which sources are polled, starting with the most important source first.

For systems with multiple interrupt sources, a flip-flop may be used to remember that an interrupt has been generated by a particular source. The microcontroller will need a sustained indicator (sometimes called a flag) to find the source of the interrupt (by reading an input port line). When the microcontroller finds the source, it can clear the flip-flop (via an output port line), to make it ready for the next interrupt.

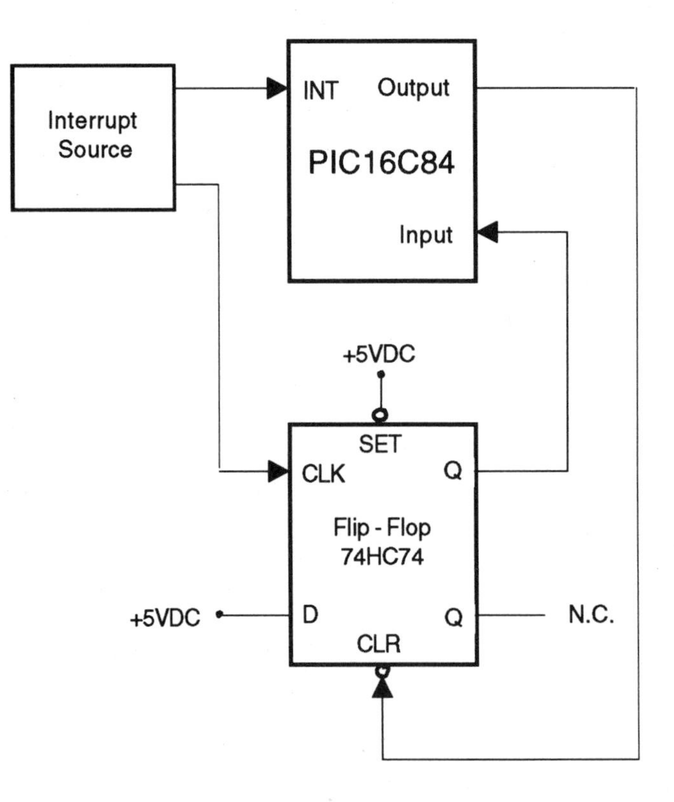

One flip - flop is required for each interrupt source. Used as a flag.

EXAMPLE - EXTERNAL INTERRUPT

INT is edge sensitive. It will respond to a rising edge if the INTEDG bit (bit 6 in the option register) is set, or to a falling edge if the INTEDG bit is clear.

To generate an INT interrupt, the INT pin must detect the edge of a pulse. For demonstration purposes, this can be done using a momentary contact toggle switch with a debouncing circuit followed by a half-monostable (one-shot) circuit which generates a negative-going pulse of 10 or more microseconds duration.

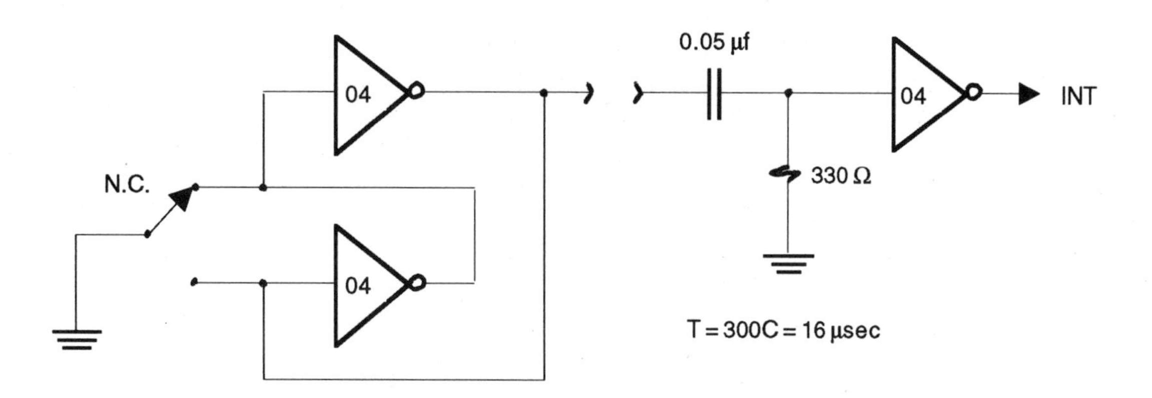

The pulser output is normally logic 1 and the output is a negative-going pulse. We will set up to respond to the falling edge.

When an INT interrupt occurs, the microcontroller:

1) Completes execution of the current instruction.
2) Tests the global interrupt enable flag. If the flag is set, interrupts are enabled.
3) Tests the INT interrupt enable flag. If the flag is set, INT interrupts are enabled and the microcontroller will begin the interrupt sequence.
4) If either interrupt enable flag indicates "disabled", the microcontroller will continue whatever it was doing when the interrupt signal was received and ignore the interrupt.
5) Clears the global interrupt enable flag to prevent further interrupts (automatic on recognition of interrupt).
6) Saves the address of the next instruction on the stack (automatic).
6) Jumps to the address pointed to by the INT vector = 0x004.

The following test program used in conjunction with the pulse circuit described above will serve to demonstrate an INT interrupt.

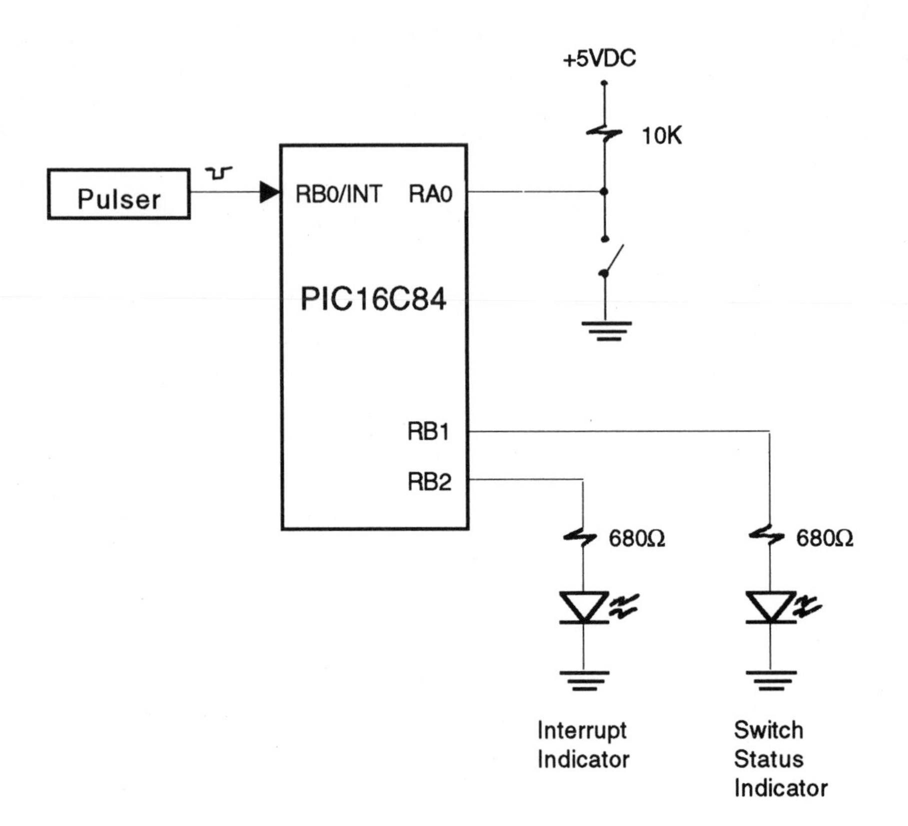

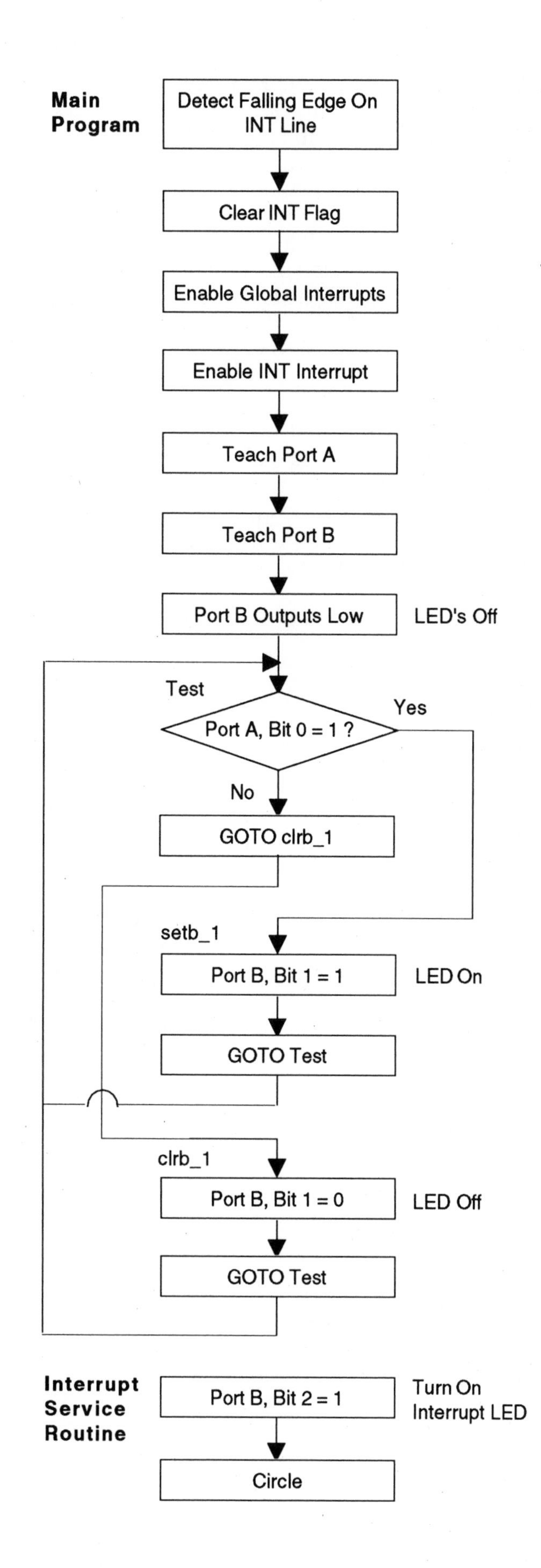
Main Program
Detect Falling Edge On INT Line
Clear INT Flag
Enable Global Interrupts
Enable INT Interrupt
Teach Port A
Teach Port B
Port B Outputs Low
LED's Off
Test
Port A, Bit 0 = 1 ?
Yes
No
GOTO clrb_1
setb_1
Port B, Bit 1 = 1
LED On
GOTO Test
clrb_1
Port B, Bit 1 = 0
LED Off
GOTO Test
Interrupt Service Routine
Port B, Bit 2 = 1
Turn On Interrupt LED
Circle

The program scans the status of the switch connected to port A, bit 0 and displays its status at port B, bit 1. Its operation serves as something to do for demonstration purposes while waiting for the interrupt to occur. The output is visible, so operation of the program can be verified by changing switch settings while the program runs. Note that the INT interrupt is enabled (essential).

If an INT interrupt occurs, the microcontroller jumps to 0x004 where the interrupt service routine begins. The interrupt service routine turns on the interrupt indicator LED at port B, bit 2 indicating an interrupt has occurred and then sits in a loop (circle).

The port B pullups are turned off. Port B, bit 0 must be an input to receive the interrupt. The program calls for response to a falling edge on the INT line.

Program an PIC16C84 with both programs. Power-up your test circuit. Change the position of the switch on port A, bit 0 to confirm that the main program is running. Pulse the INT line and observe the result at the LED connected to port B, bit 2.

The last instruction in an interrupt service routine must be a RETFIE (return from interrupt). This sends the processor back to the main program.

The register contents are not saved by the interrupt service routine in this example because the data is not needed.

Remember that an interrupt should not occur during a timing loop as it will lengthen the loop by the time required to service that interrupt. Also, an interrupt which occurs while the PIC16C84's timer is in use may or may not be serviced before the timer times out.

The use of interrupts greatly enhances the power of the microcontroller. Interrupts may be periodic, as determined by a real time clock, or may be related to an event such as a counter counting down to 0 or a burglar tripping an alarm. The microcontroller does not have to go around and around in a loop waiting for these things to happen, so it can perform other useful tasks in the meantime.

```
;=======PICT15.ASM===========================6/27/96==
;INT interrupt demo
;-----------------------------------------------------
        list    p=16c84
        radix   hex
;-----------------------------------------------------
;       destination designator equates
w       equ     0
f       equ     1
;-----------------------------------------------------
;       cpu equates (memory map)
porta   equ     0x05
portb   equ     0x06
intcon  equ     0x0b
;-----------------------------------------------------
        org     0x000
        goto    start   ;skip over location pointed
                        ;   to by interrupt vector
        org     0x004
        goto    iserv
;
start   movlw   b'01111111' ;falling edge
        option
        bcf     intcon,1  ;clear INT flag
        bsf     intcon,7  ;enable global interrupts
        bsf     intcon,4  ;enable INT interrupts
        movlw   0xff
        tris    porta   ;teach port A all inputs
        movlw   b'00000001'  ;teach port B
        tris    portb   ;bit 0 INT, rest outputs
        clrf    portb   ;port B, bits 7-1 low
test    btfss   porta,0 ;test port A, bit 0
        goto    clrb_1
setb_1  bsf     portb,1 ;set port B, bit 1
        goto    test
;
clrb_1  bcf     portb,1 ;clear port B,bit 1
        goto    test
;
iserv   bsf     portb,2 ;turn on LED
circle  goto    circle  ;done
;
        end


;-----------------------------------------------------
;at blast time, select:
;       memory unprotected
;       watchdog timer disabled (default is enabled)
;       standard crystal (using 4 MHz osc for test)
;       power-up timer on
;=====================================================
```

TIMING AND COUNTING

DIGITAL OUTPUT WAVEFORMS

Digital output waveforms are easy to generate by writing 1's and 0's to a port line. A positive going pulse of short or long duration may be output by initializing the line at 0, outputting a 1, using a software timing loop or hardware timer to measure the pulse duration, and then writing a 0 to the port line.

Square waves are easy to generate as you know from some of the examples:

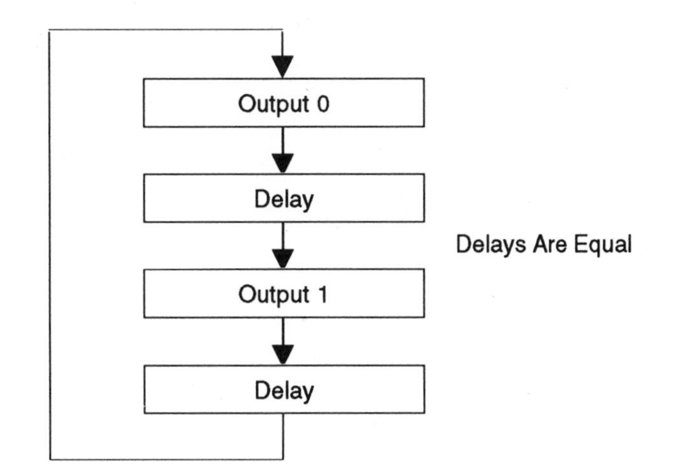

A rectangular wave is produced if the delays are not equal.

If the delays are changed each time around the loop, sweep frequencies or frequency-modulated signals may be generated.

As an alternative, the PIC16C84 timer/counter may be used. An advantage is that the microcontroller is not tied up generating repetitive waveforms.

TIMING AND COUNTING
USING THE PIC16C84'S ON-BOARD TIMER/COUNTER

The PIC16C84 timer/counter is referred to as the TIMER0 (TMR0) module. I will simply refer to it as the timer/counter.

The timer/counter's features are:

- 8-bit
- Read/write
- 8-bit software programmable prescaler
- Internal or external clock
- Edge-rising or falling (external clock)
- Increments
- Interrupt on overflow from 0xFF to 0x00 with flag output

The PIC16C84 timer/counter has an interrupt on overflow from 0xFF to 0x00 and is capable of doing other tasks while timing/counting is going on.

The option register is associated with both the timer/counter and the watchdog timer. It is in hardware (i.e., not a file register)

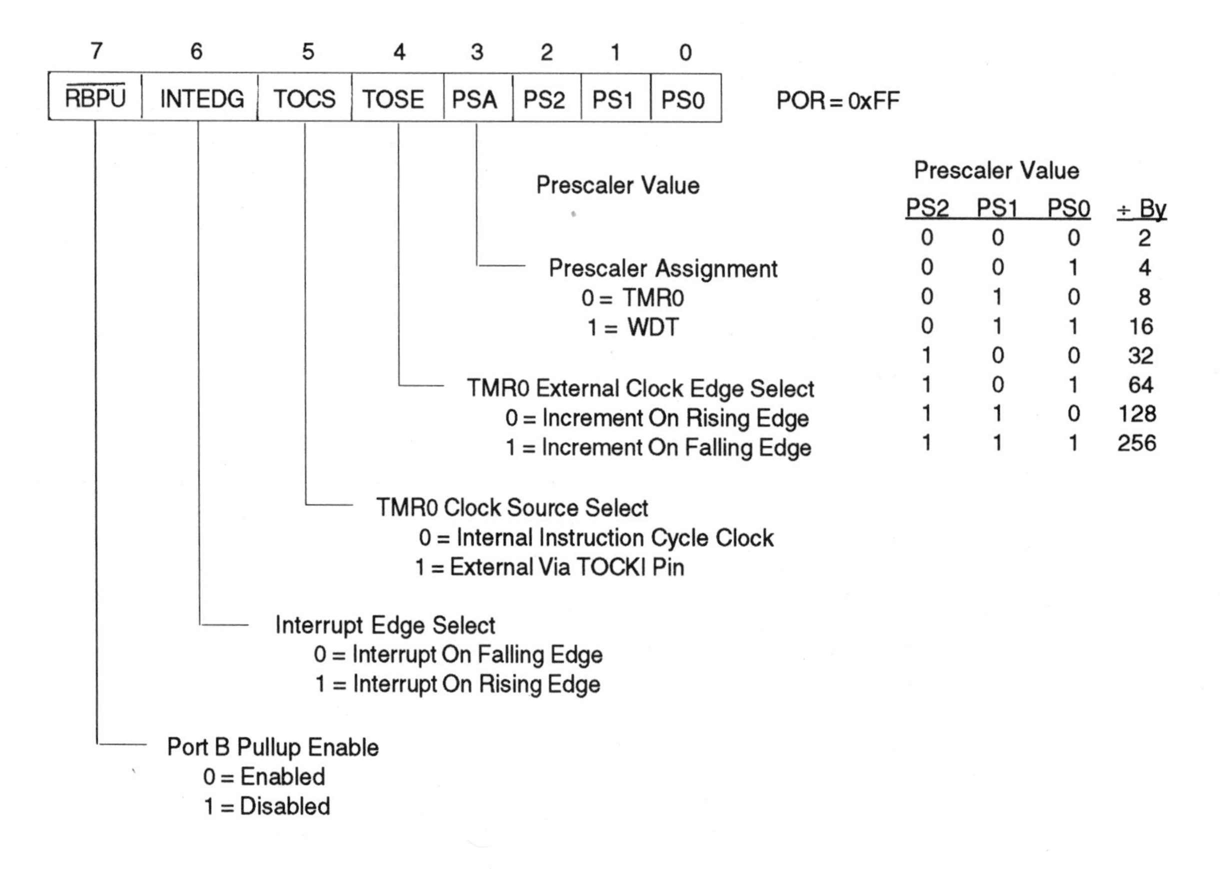

Prescaler Value

PS2	PS1	PS0	÷ By
0	0	0	2
0	0	1	4
0	1	0	8
0	1	1	16
1	0	0	32
1	0	1	64
1	1	0	128
1	1	1	256

The option register is written to by executing the OPTION instruction causing the contents of the W register to be transferred into the option register. There is another method which requires file register bank switching which we will disregard for now.

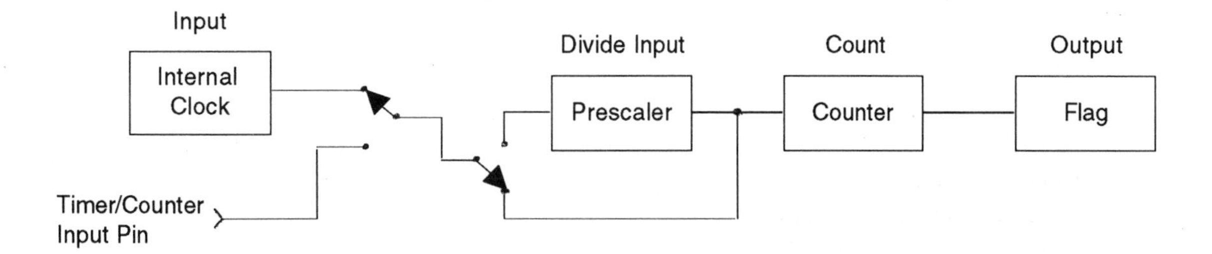

The clock source may be either the PIC16C84's internal instruction cycle clock or the T0CKI pin. An external clock may be an oscillator running (much) slower than the PIC16C84's clock oscillator or it might be a source of pulses to be counted.

The input is either fed directly to the timer/counter (by bypassing the prescaler) or through an 8-bit software programmable prescaler.

The prescaler is assigned to either the timer/counter or the watchdog timer under software control by using the OPTION instruction. Details later.

The prescaler value may be 1-of-8 as determined by the bits in the option register bits 2,1,0 (see table). Timing/counting may be done without the prescaler by assigning the prescaler to the watchdog timer (the only way to do it).

When the prescaler is assigned to the timer/counter, all instructions which write to the timer/counter register f1 (0x01) will clear the prescaler.

CLRF
MOVWF
BSF
Etc.

If an external clock source is used with no prescaler, synchronization of the external clock input must take place. This requires a short sampling procedure plus a delay after synchronization occurs and prior to the timer/counter being incremented.

There are some requirements for an external clock signal.

No Prescaler

Input high for at least 2 Tosc
Input low for at least 2 Tosc

With Prescaler

Input period of at least 4 Tosc divided by the prescaler value
Highs and lows must be of greater than 10 nanoseconds duration

Tosc is the period of the PIC16C84 clock oscillator.

If there is a write to the timer/counter, incrementing is inhibited for the next 2 instruction cycles. This can be compensated for by adjusting the number loaded in the timer/counter.

External clock pulses may be detected on their rising or falling edge (software selectable via bit 6 in the option register).

The timer/counter outputs are:

- Reading the timer/counter register (0x01).
- Interrupt on overflow from 0xFF to 0x00.

The timer is incremented by incoming pulses. When the count climbs through 0xFF, the count starts over at 0x00. The timer may be incremented over and over if need be and the number of times the counter reaches a certain value may be counted using a file register as a counter.

PRESCALER

There is an 8-bit counter which may be used either as a prescaler for the timer/counter or as a post-scaler for the watchdog timer. This counter is simply referred to as the prescaler instead of worrying about "pre" vs. "post".

The prescaler divides the clock input by one-of-eight values which effectively reduces the frequency of the clock. The prescaler is used to divide the input by:

```
  1  (bypass scaler by assigning it to the watchdog timer)
  2
  4
  8
 16
 32
 64
128
256
```

The only way to feed the clock input directly to the timer/counter is to get the prescaler out of the way by assigning it to the watchdog timer. The watchdog timer is not given much attention in this book but we at least have to be able to properly assign the prescaler to it.

The prescaler assignment and ratio are determined by 5 bits in the option register.
When the prescaler is assigned to TMR0, all instructions which write to TMR0 such as CLRF, MOVWF, BSF, etc. will clear the prescaler to prepare it for division of the input signal.

Changing Prescaler Assignment

From TMR0 To WDT

```
clrf      tmr0          ;clear timer/counter
clrwdt                  ;clear watchdog timer
movlw     b'xxxx1xxx'   ;prescaler assign TMR0
option                  ;   new prescale value
```

From WDT To TMR0

```
clrwdt                  ;clear WDT and prescaler
movlw     b'xxxx0xxx'   ;select TMR0, prescale value
option                  ;   and clock source
```

This sequence must be used even if the watchdog timer is disabled.

The CLRWDT instruction should precede switching prescaler assignment.

USING THE TIMER/COUNTER

Setting up the timer/counter

- Assign prescaler per procedure above.
- Set up the timer/counter by sending the proper bits to option register

Starting the timer/counter

Write a number to the timer/counter file register

Counter

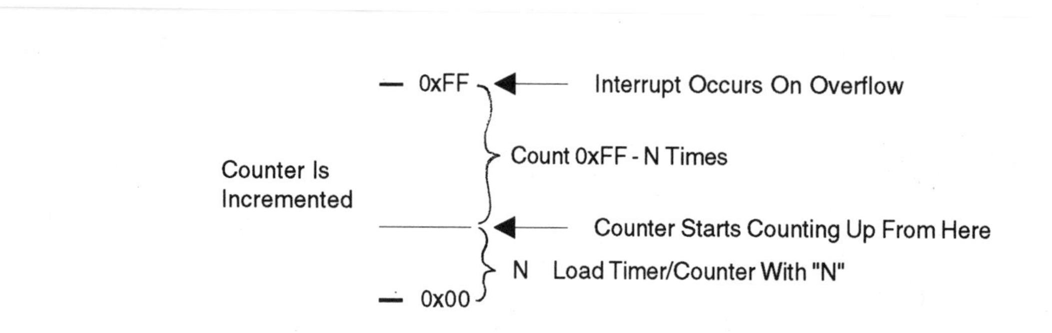

How do we know the timer/counter is doing something?

- Successive reads of the timer/counter file register or test a bit in the timer/counter file register
- Interrupt on overflow 0xFF to 0x00. This is how we know the timer/counter is finished counting. The PIC16C84 is free to do other things while the timer/counter is doing it's thing. The interrupt flag is the output.

Timer/counter will keep counting as long as:

- It is not cleared or written to by program instructions
- The microcontroller is not reset

Timer/counter must be reloaded after each overflow for repeating time intervals

If this is not done, the count will start at 0x00 each time.

Stopping the timer/counter

Can't - it just runs

Experiments follow which will illustrate the use of the timer/counter.

TIMER/COUNTER EXPERIMENTS

Digital Output Waveform Using TMR0 - Internal Clock

Use internal clock divided by 256
Blink an LED at fast rate - delay 8.4 milliseconds

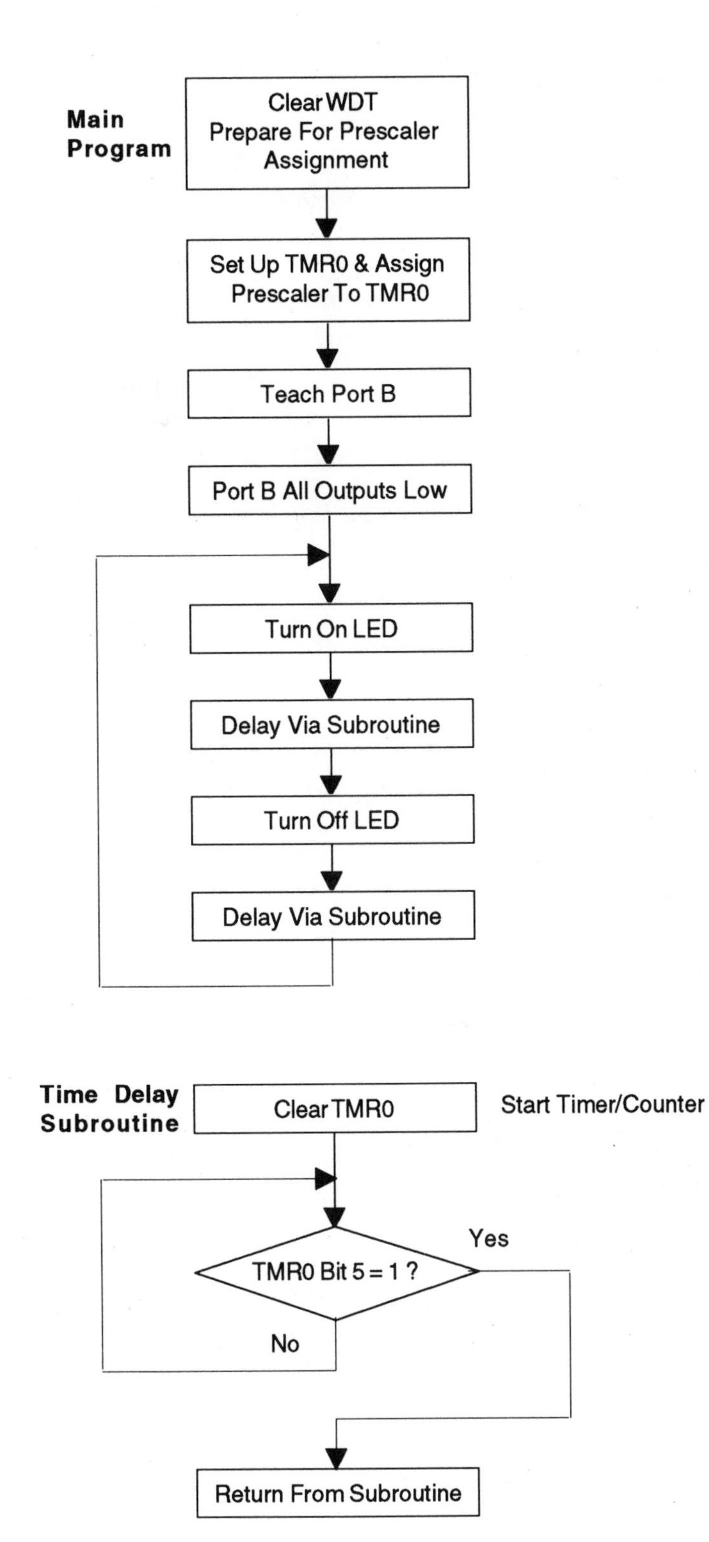

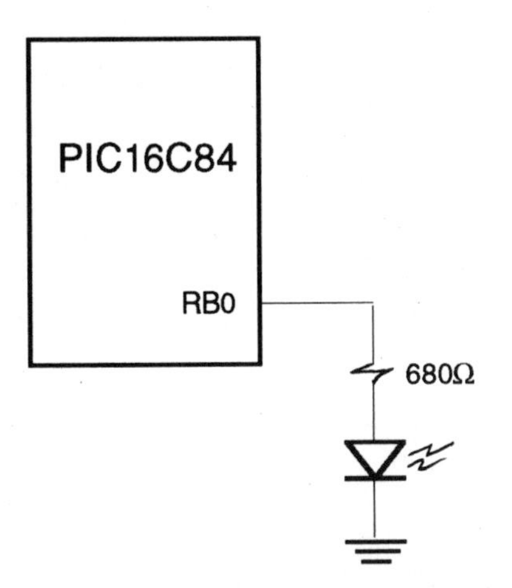

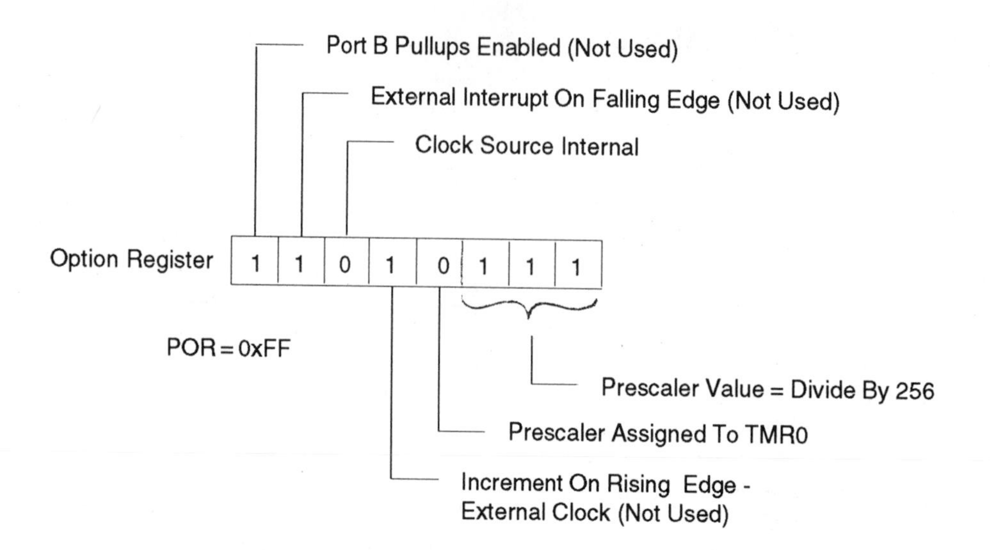

The timer/counter may be read as a whole or a single bit may be tested as is done in this example. When bit 5 of the timer/counter increments to "1", the count is 32. This works for numbers which are a power of 2.

The delay is roughly 256 microseconds per pulse into TMR0 times 32 (via TMR0 bit 5) = 8.2 milliseconds.

```
;=======PICT7.ASM============================6/25/96==
;time delay demo
;       tmr0, internal clock divided by 256
;       watch tmr0, bit 5 = count to 32
;       delay = 8.2 milliseconds
;---------------------------------------------------
        list    p=16c84
        radix   hex
;---------------------------------------------------
;       destination designator equates
w       equ     0
f       equ     1
;---------------------------------------------------
;       cpu equates (memory map)
tmr0    equ     0x01
portb   equ     0x06
count   equ     0x0c
;---------------------------------------------------
        org     0x000
;
start   clrwdt          ;prep for assign prescaler
        movlw   b'11010111' ;assign prescaler,
;                       internal clock, divide by 256
        option
        movlw   0x00    ;load w with 0x00
        tris    portb   ;copy w tristate, port B
;                          outputs
```

```
        clrf    portb   ;all lines low
go      bsf     portb,0 ;turn on LED
        call    delay   ;delay via sub
        bcf     portb,0 ;turn off LED
        call    delay   ;delay via sub
        goto    go      ;repeat
;
delay   clrf    tmr0    ;clear TMR0, start counting
again   btfss   tmr0,5  ;bit 5 set ?
        goto    again   ;no, clear, again
        return          ;yes, end delay
;
        end
;-----------------------------------------------------
;at blast time, select:
;       memory unprotected
;       watchdog timer disabled (default is enabled)
;       standard crystal (using 4 MHz osc for test)
;       power-up timer on
;=====================================================
```

Single Time Interval - Internal Clock

Use internal clock divided by 128 and file register counter
Blink an LED once

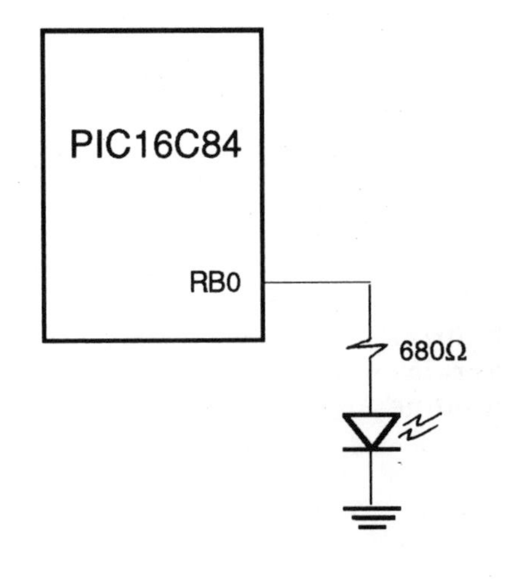

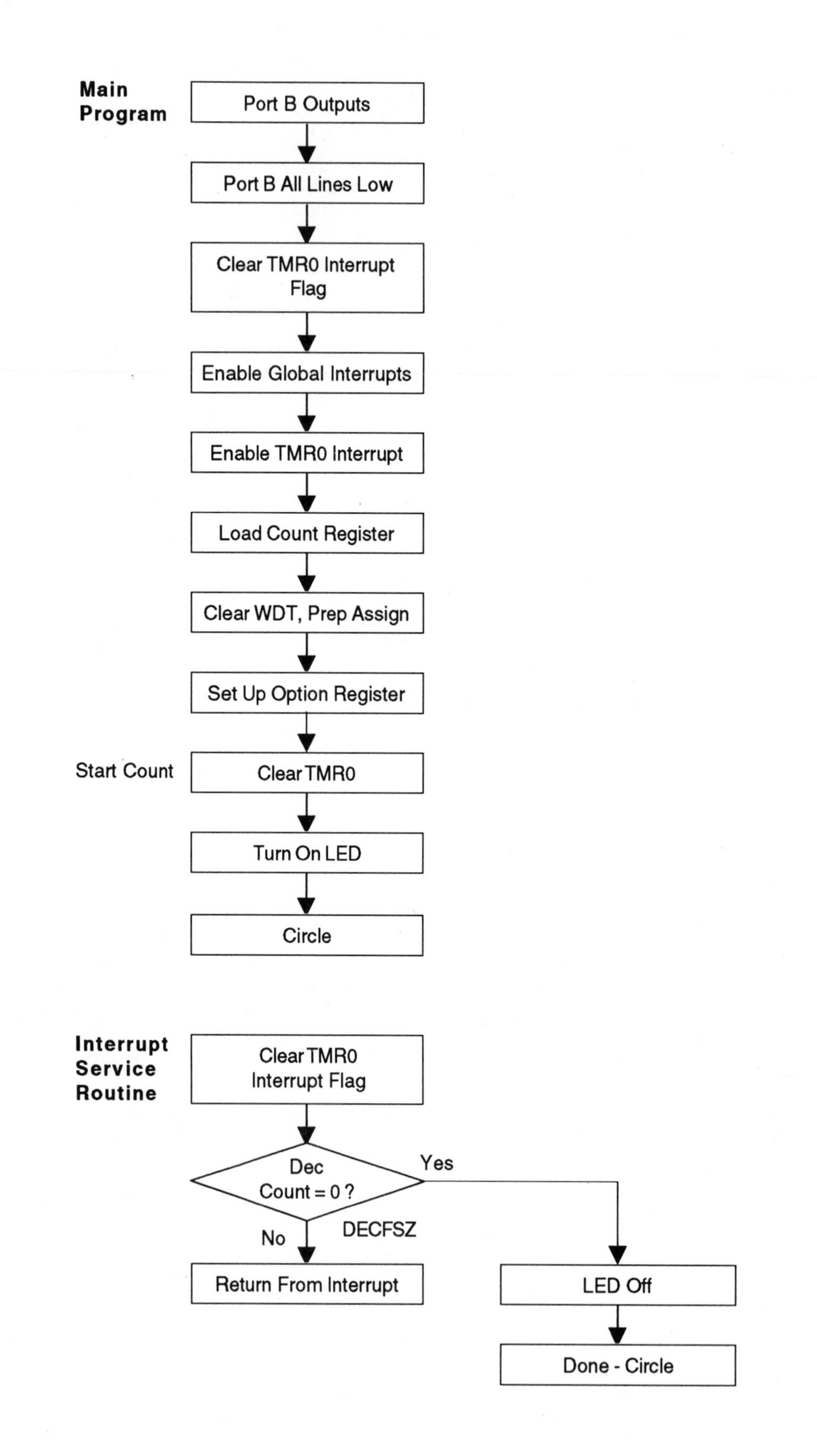

Main Program
Port B Outputs
Port B All Lines Low
Clear TMR0 Interrupt Flag
Enable Global Interrupts
Enable TMR0 Interrupt
Load Count Register
Clear WDT, Prep Assign
Set Up Option Register
Start Count
Clear TMR0
Turn On LED
Circle
Interrupt Service Routine
Clear TMR0 Interrupt Flag
Dec Count = 0 ?
Yes
No
DECFSZ
Return From Interrupt
LED Off
Done - Circle

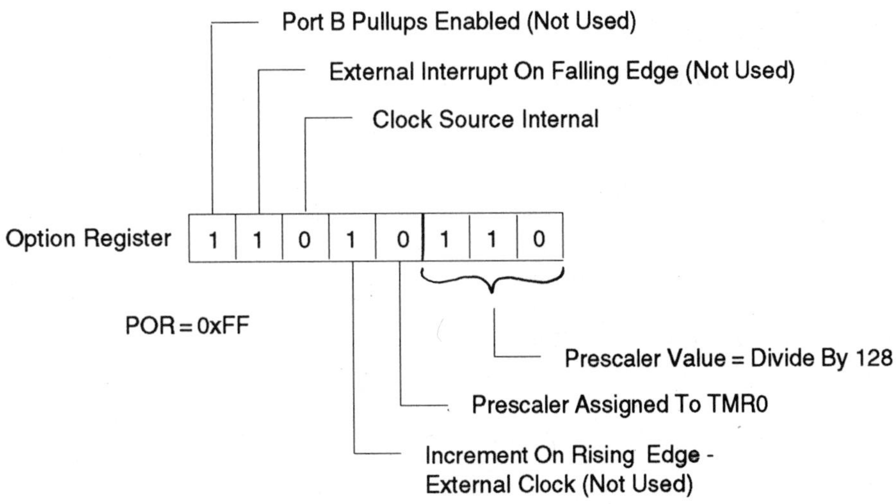

```
;=======PICT16.ASM==========================6/28/96==
;timer/counter demo
;       single time interval
;       internal clock divided by 128
;       file register counter
;---------------------------------------------------
        list    p=16c84
        radix   hex
;---------------------------------------------------
;       destination designator equates
w       equ     0
f       equ     1
;---------------------------------------------------
;       cpu equates (memory map)
tmr0    equ     0x01
portb   equ     0x06
intcon  equ     0x0b
count   equ     0x0c
;---------------------------------------------------
        org     0x000
        goto    start   ;skip over location pointed
                        ;   to by interrupt vector
        org     0x004
        goto    iserv
;
start   movlw   0x00    ;port B outputs
        tris    portb
        clrf    portb   ;all lines low (bit 0 LED off)
        bcf     intcon,2  ;clear TMR0 interrupt flag
        bsf     intcon,7  ;enable global interrupts
        bsf     intcon,5  ;enable TMR0 interrupts
        movlw   0xff    ;decimal 256
        movwf   count   ;load counter
;
        clrwdt          ;clr WDT prep prescale assign
        movlw   b'11010110'  ;set up timer/counter
```

```
        option
;
        clrf    tmr0    ;start timer/c, clr prescaler
        bsf     portb,0 ;turn on LED
circle  goto    circle  ;wait for interrupt
;
iserv   bcf     intcon,2  ;clear TMR0 interrupt flag
        decfsz  count,f ;count = 0?
        retfie          ;no
        bcf     portb,0 ;yes, LED off
round   goto    round   ;done, circle
;
        end
;------------------------------------------------------
;at blast time, select:
;       memory unprotected
;       watchdog timer disabled (default is enabled)
;       standard crystal (using 4 MHz osc for test)
;       power-up timer on
;======================================================
```

Single Time Interval - External Clock

Use external 0.1 second clock (555 1-shot or another PIC16/17)
Bypass prescaler
Blink an LED once

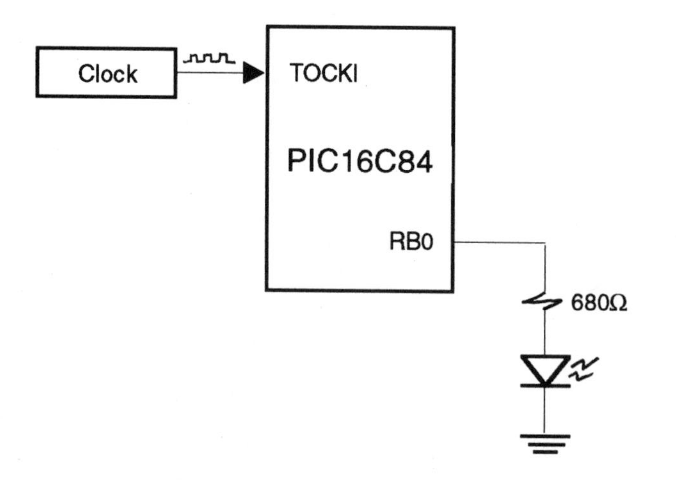

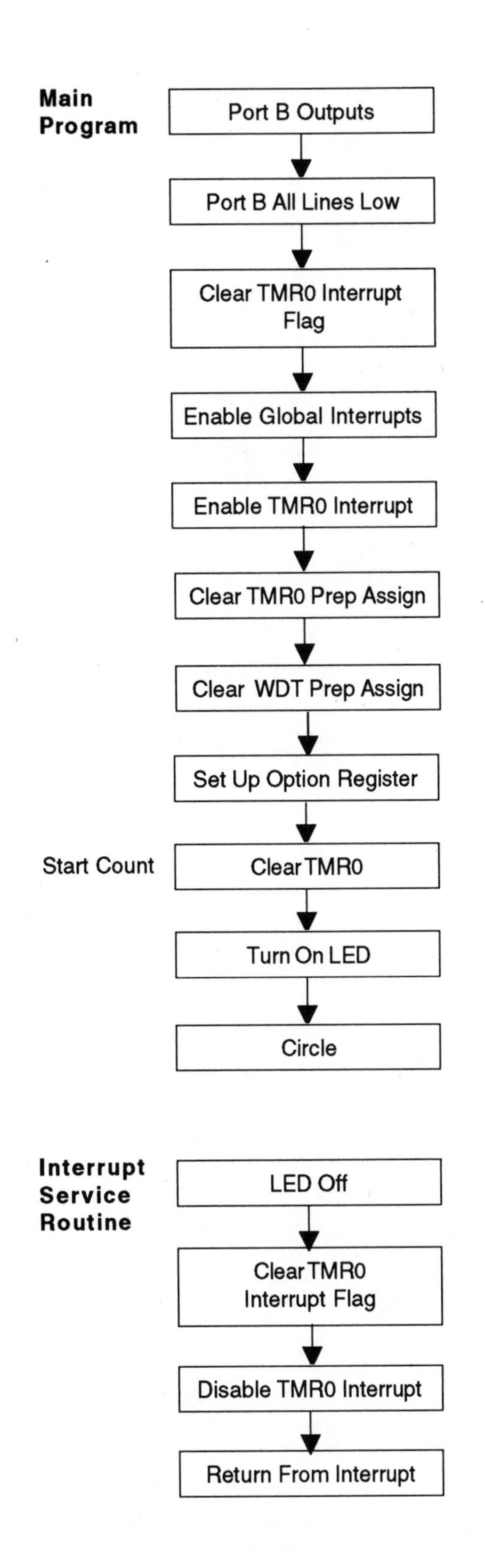
Main Program
Port B Outputs
Port B All Lines Low
Clear TMR0 Interrupt Flag
Enable Global Interrupts
Enable TMR0 Interrupt
Clear TMR0 Prep Assign
Clear WDT Prep Assign
Set Up Option Register
Start Count
Clear TMR0
Turn On LED
Circle
Interrupt Service Routine
LED Off
Clear TMR0 Interrupt Flag
Disable TMR0 Interrupt
Return From Interrupt

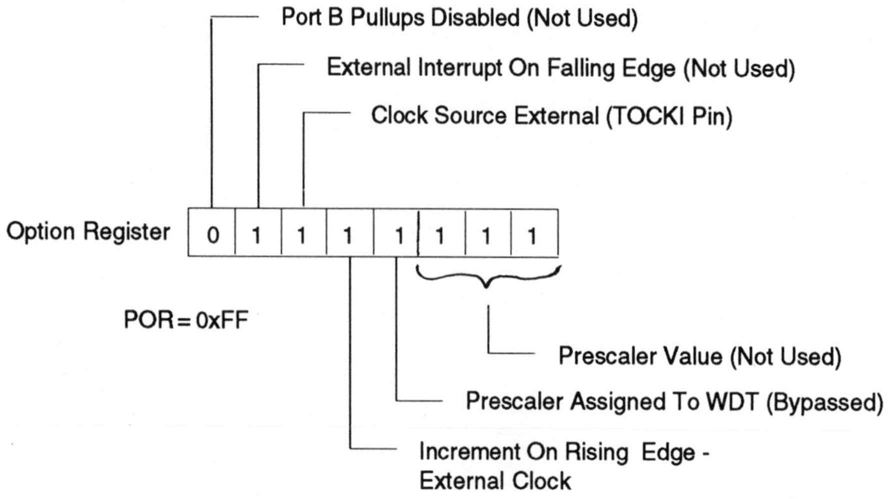

Port B Pullups Disabled (Not Used)
External Interrupt On Falling Edge (Not Used)
Clock Source External (TOCKI Pin)

Option Register	0	1	1	1	1	1	1	1

POR = 0xFF

Prescaler Value (Not Used)
Prescaler Assigned To WDT (Bypassed)
Increment On Rising Edge - External Clock

```
;=======PICT17.ASM===========================6/28/96==
;timer/counter demo
;       single time interval
;       external clock, bypass prescaler
;       blink LED once
;-----------------------------------------------------
        list    p=16c84
        radix   hex
;-----------------------------------------------------
;       destination designator equates
w       equ     0
f       equ     1
;-----------------------------------------------------
;       cpu equates (memory map)
tmr0    equ     0x01
portb   equ     0x06
intcon  equ     0x0b
;-----------------------------------------------------
        org     0x000
        goto    start   ;skip over location pointed
                        ;   to by interrupt vector
        org     0x004
        goto    iserv
;
start   movlw   0x00    ;port B outputs
        tris    portb
        clrf    portb   ;all lines low (bit 0 LED off)
        bcf     intcon,2  ;clear TMR0 interrupt flag
        bsf     intcon,7  ;enable global interrupts
        bsf     intcon,5  ;enable TMR0 interrupts
;
        clrf    tmr0    ;clear timer/counter
        clrwdt          ;clr WDT prep prescale assign
        movlw   b'01111111'  ;set up timer/counter
        option
```

```
;
        clrf    tmr0     ;start timer/c, clr prescaler
        bsf     portb,0 ;turn on LED
circle  goto    circle  ;wait for interrupt 1st time,
;                           done 2nd time
;
iserv   bcf     portb,0 ;LED off
        bcf     intcon,2  ;clear TMR0 interrupt flag
        bcf     intcon,5  ;disable TMR0 interrupt
        retfie           ;done
;
        end
;-----------------------------------------------------
;at blast time, select:
;       memory unprotected
;       watchdog timer disabled (default is enabled)
;       standard crystal (using 4 MHz osc for test)
;       power-up timer on
;=====================================================
```

Free Running Mode - Internal Clock

Use internal clock divided by 128
Output to port B, bit 0

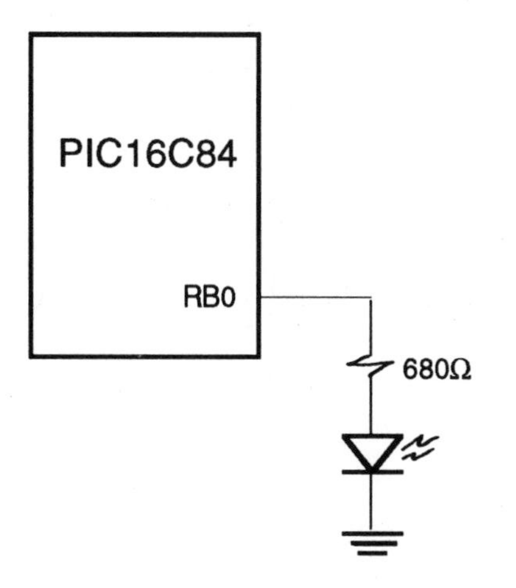

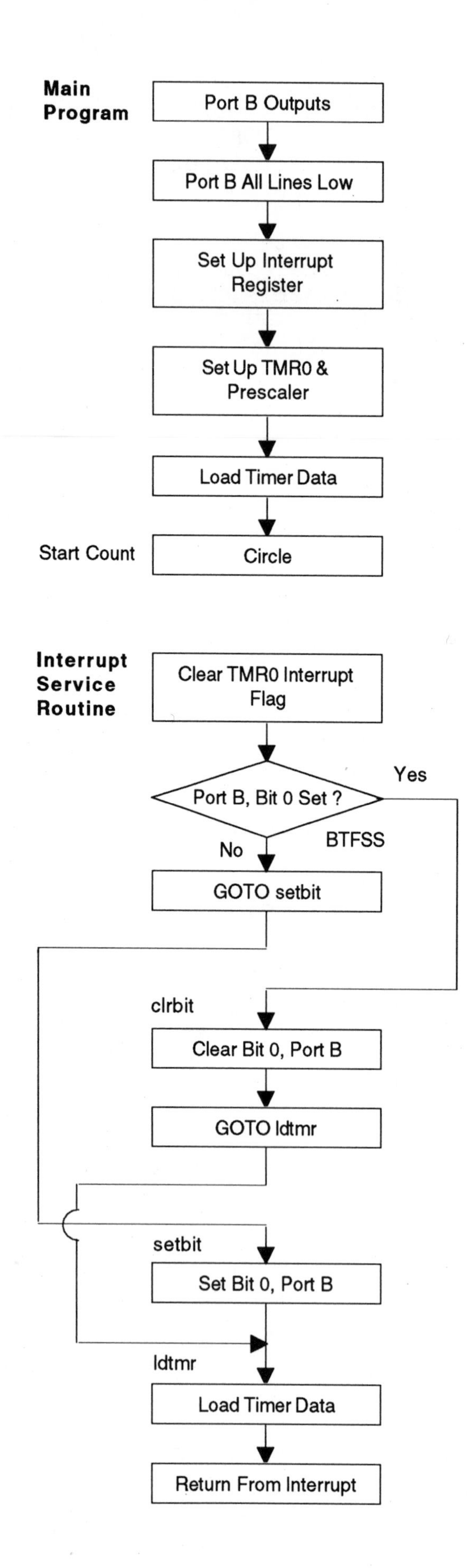

Main Program
Port B Outputs
Port B All Lines Low
Set Up Interrupt Register
Set Up TMR0 & Prescaler
Load Timer Data
Start Count
Circle
Interrupt Service Routine
Clear TMR0 Interrupt Flag
Port B, Bit 0 Set ?
Yes
No
BTFSS
GOTO setbit
clrbit
Clear Bit 0, Port B
GOTO ldtmr
setbit
Set Bit 0, Port B
ldtmr
Load Timer Data
Return From Interrupt

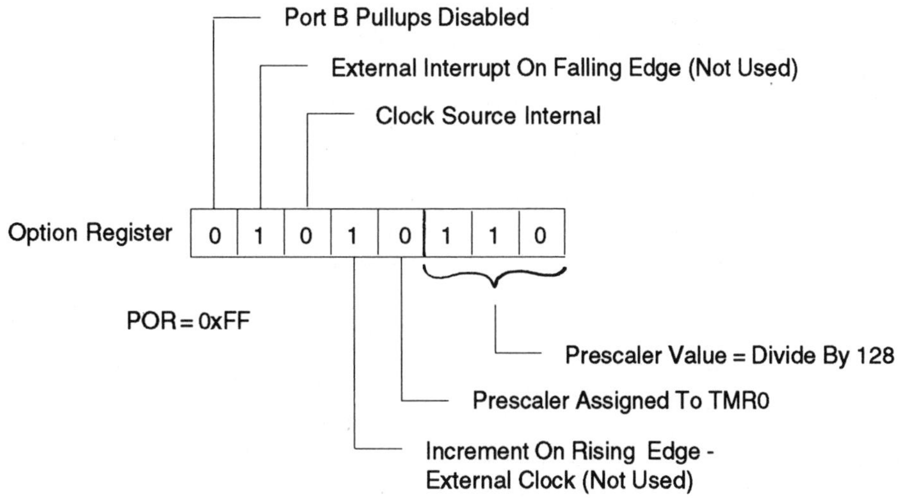

Port B Pullups Disabled

External Interrupt On Falling Edge (Not Used)

Clock Source Internal

Option Register

0	1	0	1	0	1	1	0

POR = 0xFF

Prescaler Value = Divide By 128

Prescaler Assigned To TMR0

Increment On Rising Edge -
External Clock (Not Used)

```
;=======PICT20.ASM===========================6/28/96==
;timer/counter demo
;       free running
;       internal clock divided by 128
;       10 counts to overflow
;---------------------------------------------------
        list    p=16c84
        radix   hex
;---------------------------------------------------
;       destination designator equates
w       equ     0
f       equ     1
;---------------------------------------------------
;       cpu equates (memory map)
tmr0    equ     0x01
portb   equ     0x06
intcon  equ     0x0b
;---------------------------------------------------
        org     0x000
        goto    start   ;skip over location pointed
                        ;   to by interrupt vector
        org     0x004
        goto    iserv
;
start   movlw   0x00    ;port B outputs
        tris    portb
        clrf    portb   ;all lines low (bit 0 LED off)
        bcf     intcon,2  ;clear TMR0 interrupt flag
        bsf     intcon,7  ;enable global interrupts
        bsf     intcon,5  ;enable TMR0 interrupts
;
        clrwdt          ;clr WDT prep prescale assign
        movlw   b'01010110'  ;set up timer/counter
        option
;
        movlw   0xf5    ;10 counts (decimal)
```

```
        movwf   tmr0    ;start timer/c, clr prescaler
circle  goto    circle  ;wait for interrupt
;
iserv   bcf     intcon,2  ;clear TMR0 interrupt flag,
;                            enable further interrupts
        btfss   portb,0   ;port B, bit 0 status?
        goto    setbit  ;bit is clear
clrbit  bcf     portb,0 ;clear port B, bit 0
        goto    ldtmr   ;to load timer/counter
setbit  bsf     portb,0 ;set port B, bit 0
ldtmr   movlw   0xf5    ;10 counts (decimal)
        movwf   tmr0    ;start timer/c, clr prescaler
        retfie          ;return from interrupt
;
        end
;------------------------------------------------------
;at blast time, select:
;       memory unprotected
;       watchdog timer disabled (default is enabled)
;       standard crystal (using 4 MHz osc for test)
;       power-up timer on
;======================================================
```

Run the program and look at port B, bit 0 with a scope.

Examples:

- Load 0xF5, prescaler ÷ 128

  1µsec x 128/count x 10 = 1.28 msec

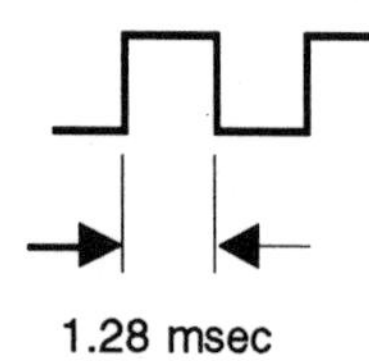

This is the time between each HI/LO or LO/HI transition at the port line. The time to execute the interrupt service routine adds to this slightly.

- Load 0x00, prescaler ÷ 128

  0x00 = 256 decimal   128 µsec x 256 = 33 msec
  Counts to overflow

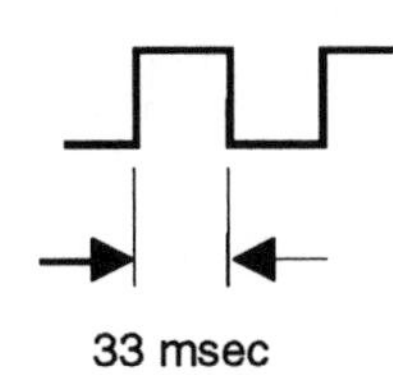

- Load 0x00, prescaler bypassed

  1 µsec x 256 = 256+ µsec (no allowance for program overhead)

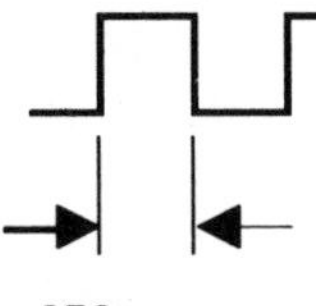
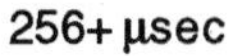

- Load 0x40, prescaler ÷ 2

  1 µsec x 192 ÷ 2 = 96+ µsec (no allowance for program overhead

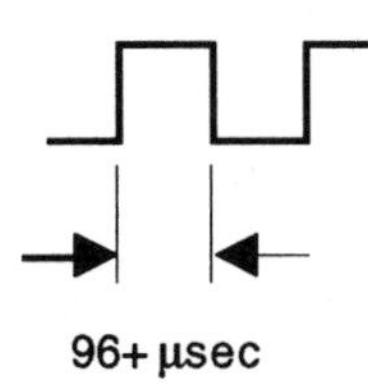

Free Running Mode - External Clock

It's your turn!

Counting Events (Pulses)

Use pulser for input.

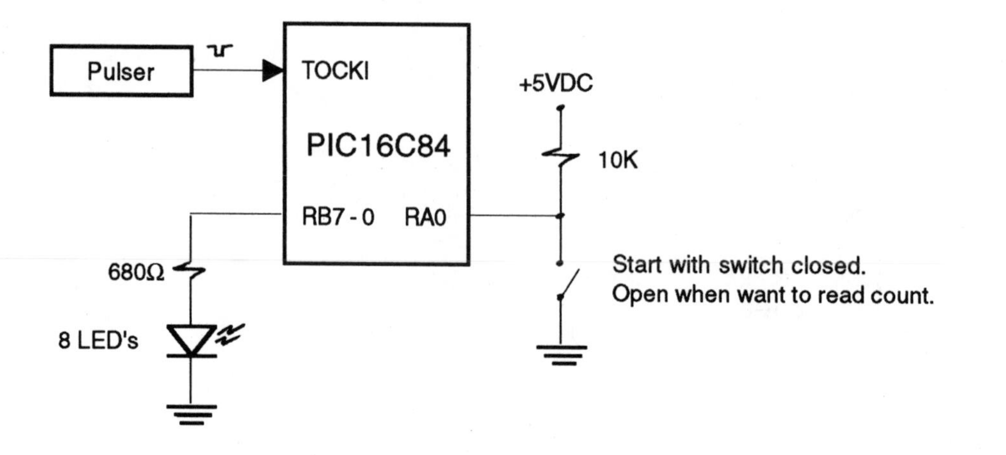

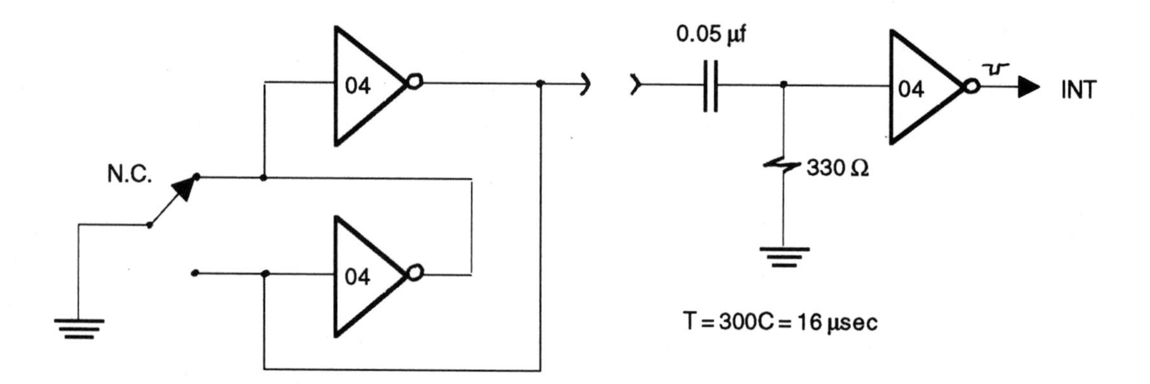

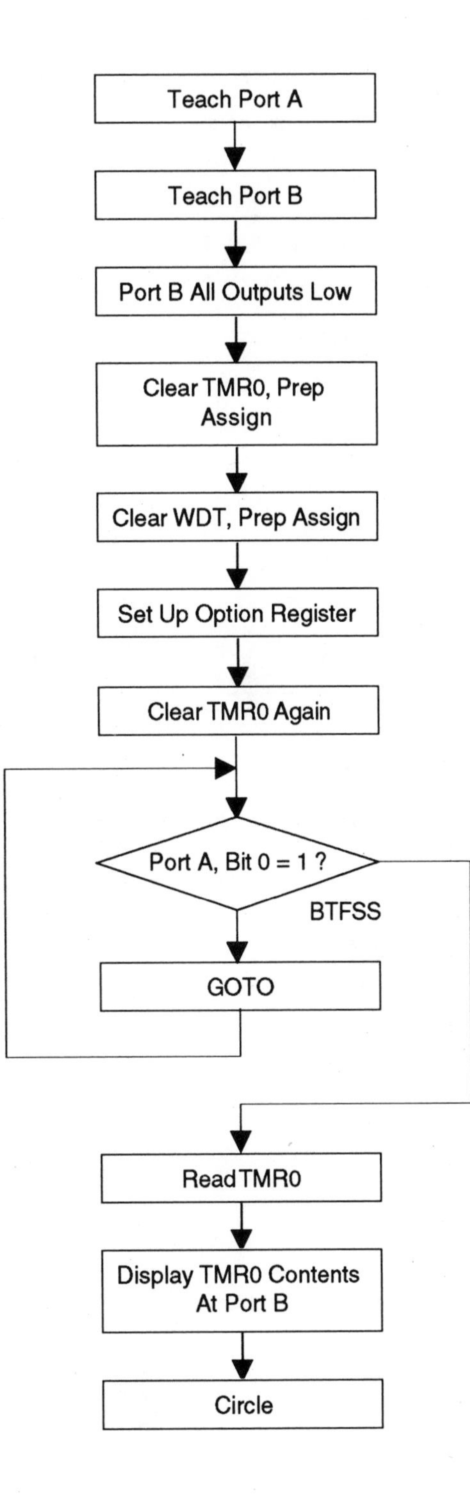
Teach Port A
Teach Port B
Port B All Outputs Low
Clear TMR0, Prep Assign
Clear WDT, Prep Assign
Set Up Option Register
Clear TMR0 Again
Port A, Bit 0 = 1 ?
BTFSS
GOTO
Read TMR0
Display TMR0 Contents At Port B
Circle

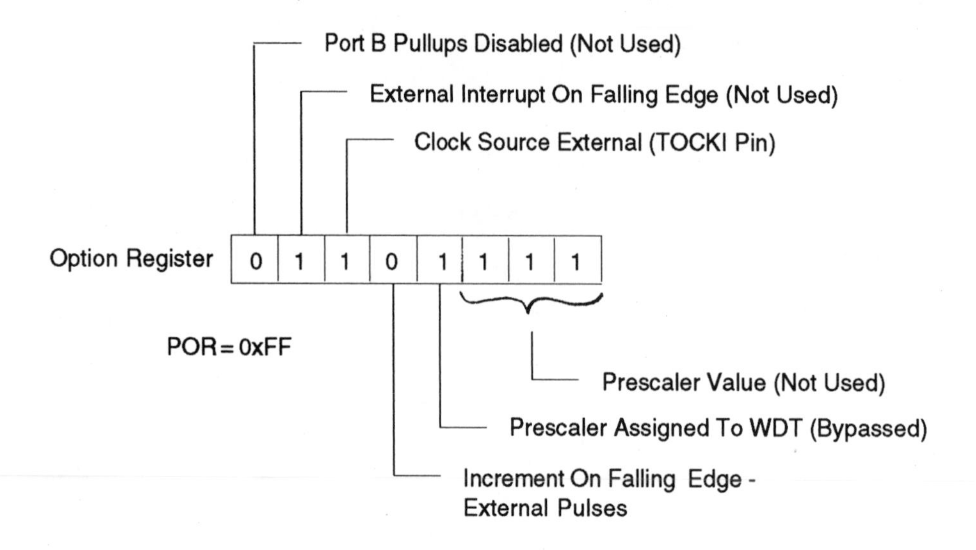


```
;=======PICT21.ASM============================6/28/96==
;event counting demo
;----------------------------------------------------
        list    p=16c84
        radix   hex
;----------------------------------------------------
;       destination designator equates
w       equ     0
f       equ     1
;----------------------------------------------------
;       cpu equates (memory map)
tmr0    equ     0x01
porta   equ     0x05
portb   equ     0x06
;----------------------------------------------------
        org     0x000
;
start   movlw   0xff      ;port A inputs
        tris    porta
        movlw   0x00      ;port B outputs
        tris    portb
        clrf    portb     ;all lines low (bit 0 LED off)
;
        clrf    tmr0      ;clear tmr0 before assign wdt
        clrwdt            ;clr WDT prep prescale assign
        movlw   b'01101111'  ;set up timer/counter
        option
        clrf    tmr0      ;clear again to zero
;
switch  btfss   porta,0 ;monitor switch
        goto    switch
        movf    tmr0,w  ;read TMR0
        movwf   portb,  ;display TMR0 contents
circle  goto    circle  ;done
```

```
;----------------------------------------------------------
;at blast time, select:
;       memory unprotected
;       watchdog timer disabled (default is enabled)
;       standard crystal (using 4 MHz osc for test)
;       power-up timer on
;==========================================================
```

1). Power-up with switch closed.

- Open switch - all LED's should be off.

2). Power-up with switch closed.

- Pulse X (few) times.
- Open switch. LED's display pulse count X.

PIC16C54

The PIC16C54 is one of the simplest and certainly the most popular member of the baseline PIC16C5X family. It is an 18-pin device. The main features, or lack of them, which differentiate the PIC16C54 from the PIC16C84 are:

- EPROM program memory.
- 512 program memory locations.
- 32 file registers. 25 are general-purpose.
- Fewer instructions (4).
- 2-level stack.
- No interrupt capability.
- No program counter high latch.
- No file register bank-switching.
- Reset vector points to 0x1FF.
- Port data-direction (TRIS) register is "buried".
- Option register is "buried".
- One timer/counter (RTCC/TMR0).
- The timer/counter RTCC/TMR0 will not interrupt the processor on overflow from 0xFF to 0x00.
- No power-up timer.

The PIC16C54 has plenty of features and is ideal for many low-end applications.

PINS AND FUNCTIONS

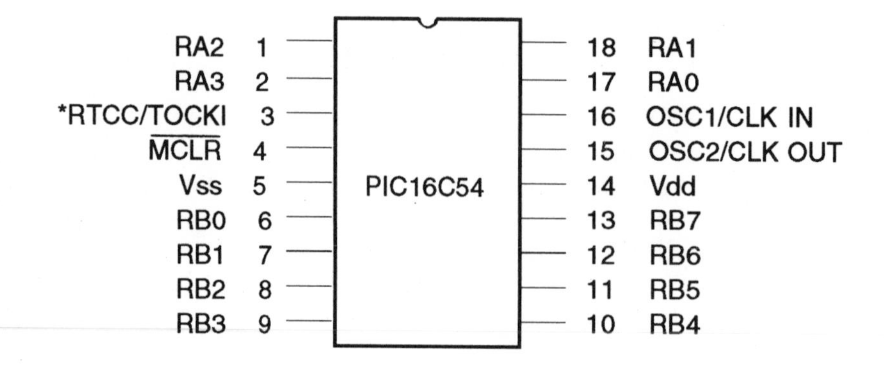

* Microchip changed pin and timer/counter names along the way. This is the real time clock counter (RTCC) input pin which is connected to what is more recently referred to as TIMER0 (TMR0) module timer/counter. Confusing. You will see both in the literature. Just think of this as the timer/counter pin!

The PIC16C54 is a CMOS device.

PACKAGES

The PIC16C54 is available in the following packages suitable for the experimenter.

Program Memory	Package	Reprogrammable
EPROM	Plastic	No, one-time programmable (OTP)
EPROM	Windowed	Yes, erase w/uv light

OTP parts contain an EPROM, but since the parts do not have windows, they cannot be erased. OTP parts cost less because they don't have windows.

CLOCK OSCILLATOR

Four different clock oscillators may be used as with the PIC16C54.

Windowed EPROM parts may be programmed to operate with any of the four clock oscillator types. This may be changed when the part is erased and reprogrammed.

For the original one-time-programmable (OTP) parts (no A suffix), each part is designed to be used with one and only one of the four oscillator types. Each part type has a unique part number (see Microchip data book). The decision of which clock oscillator to use is made prior to purchase.

The newer PIC16C54A (A suffix) OTP parts are designed so that the clock oscillator type is chosen at the time the chip is programmed. This is the primary difference between parts with and without the A suffix.

CONFIGURATION "FUSES"

The configuration "fuses" work the same as the PIC16C84.

A note about the -JW windowed parts - don't turn on the code protection bit! It can't ever be turned off. The windowed part will become an expensive OTP part.

PORTS

Port A is 4 bits/lines wide and port B is 8 bits/lines wide or byte-wide. Each port line may be individually programmed as an input line or output line. This is done using a special instruction (TRIS) which matches a bit pattern with the port lines. A 0 associated with a port line makes it an output, a 1 makes it an input. Examples follow.

The PIC16C54 does not have port data direction registers as does the PIC16C84. Use of the TRIS instruction is the way it is done with the PIC16C54. The PIC16C54 was one of Microchip's first products and this is where the TRIS instruction comes from.

See the data book for port current limit specs (less than PIC16C84).

All unused port lines should be tied to the power supply (CMOS rule - all inputs must go somewhere). On reset, all port lines are inputs.

ARCHITECTURE

Program Memory

The PIC16C54 program memory is 12 bits wide and 512 words long.

Program memory may be ROM, EPROM or OTP EPROM. Regardless of type, program memory is read-only at run time. PIC16/17's can only execute code contained in program memory.

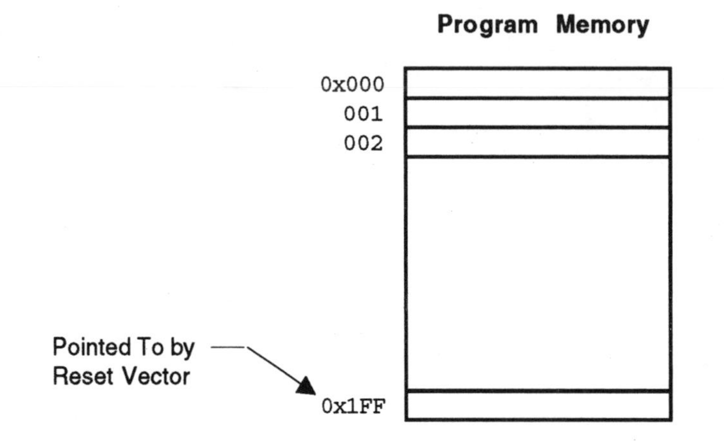

File Registers

The file registers are 8 bits wide with the exception of the program counter which is 9 bits wide. The PIC16C54 has 32 file registers (0x00 - 0x1F).

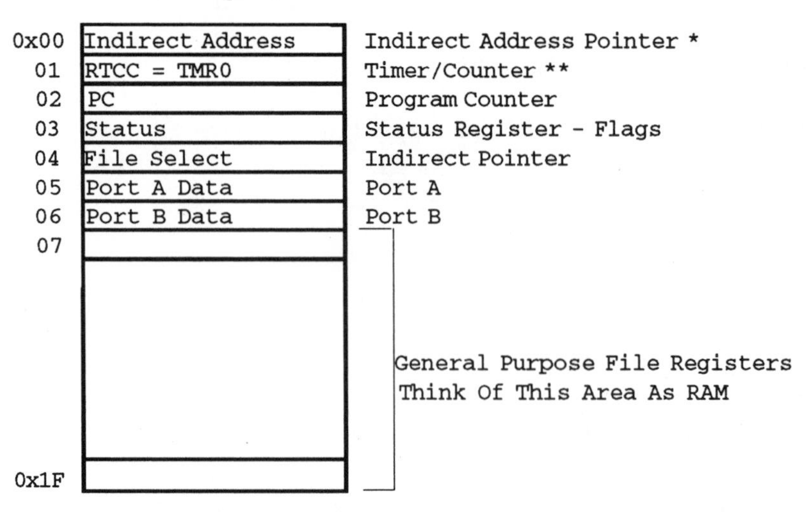

* Not Physically Implemented
** Microchip Changed Name Along The Way

The first seven file registers have specific dedicated purposes (to be described later). The remaining 26 file registers are there for your use and may be thought of as RAM or data memory for storing data during program execution.

	Hex Address	
f0	0x00	Indirect data addressing register. See indirect addressing section in programming chapter.
f1	0x01	Real time clock/counter register (RTCC or TMR0) See timing and counting chapter.
f2	0x02	Program counter. See relative addressing section in programming chapter.
f3	0x03	Status word register.
f4	0x04	File select register (FSR). See indirect addressing section in programming chapter.
f5	0x05	Port A - 4-bit, bits 4 - 7 are not implemented and read as 0's.
f6	0x06	Port B - 8-bit.
f7 - f1F	0x07 - 0x1F	General purpose registers (RAM).

Stack

The 2-level stack is in hardware meaning that it is entirely separate from the file registers (RAM) and cannot overwrite them. Subroutines can be nested 2-deep.

Reset Vector

On reset which occurs on power-up or when the reset switch is used to pull $\overline{MCLR}$ low, the PIC16C54 will look at a special memory location (0x1FF) to see where the first instruction is stored. It then goes to that address and begins executing instructions stored sequentially in memory. Generally, the address chosen for the first program instruction is 0x000 and a GOTO instruction is placed in location 0x1FF.

```
            list        p=16c54
            radix       hex
            org         0x000
start       movlw       0x00
            tris        portb
            movlw       0x0f
            movwf       portb
circle      goto        circle
            org         0x1ff
            goto        start
            end
```

Note the second ORG statement and code telling the PIC16C54 to go to "start" which is at 0x000.

Program Counter

The program counter (PC - at 0x02) is 9 bits wide. Bit 8 (9th bit) of the PC is cleared by:

- A CALL instruction
- Any instruction which writes to the PC

A GOTO loads all 9 bits of the program counter and will jump to anywhere in the PIC16C54's program memory space.

A CALL loads the lower 8 bits of the program counter. The 9th bit is cleared to 0. All subroutine calls are limited to the first 256 locations of program memory for the PIC16C54.

Any instruction which writes to the program counter changes the low 8 bits and clears the 9th bit, so computed jumps are limited to the first 256 locations of program memory for the PIC16C54.

Option Register

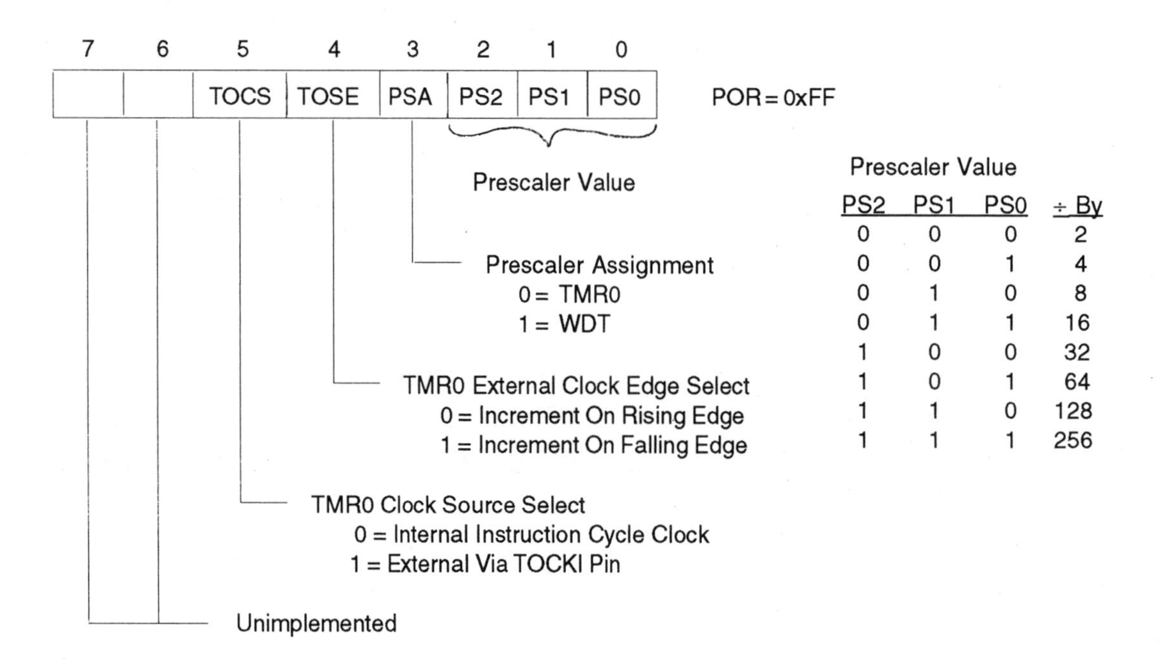

PS2	PS1	PS0	÷ By
0	0	0	2
0	0	1	4
0	1	0	8
0	1	1	16
1	0	0	32
1	0	1	64
1	1	0	128
1	1	1	256

Status Register

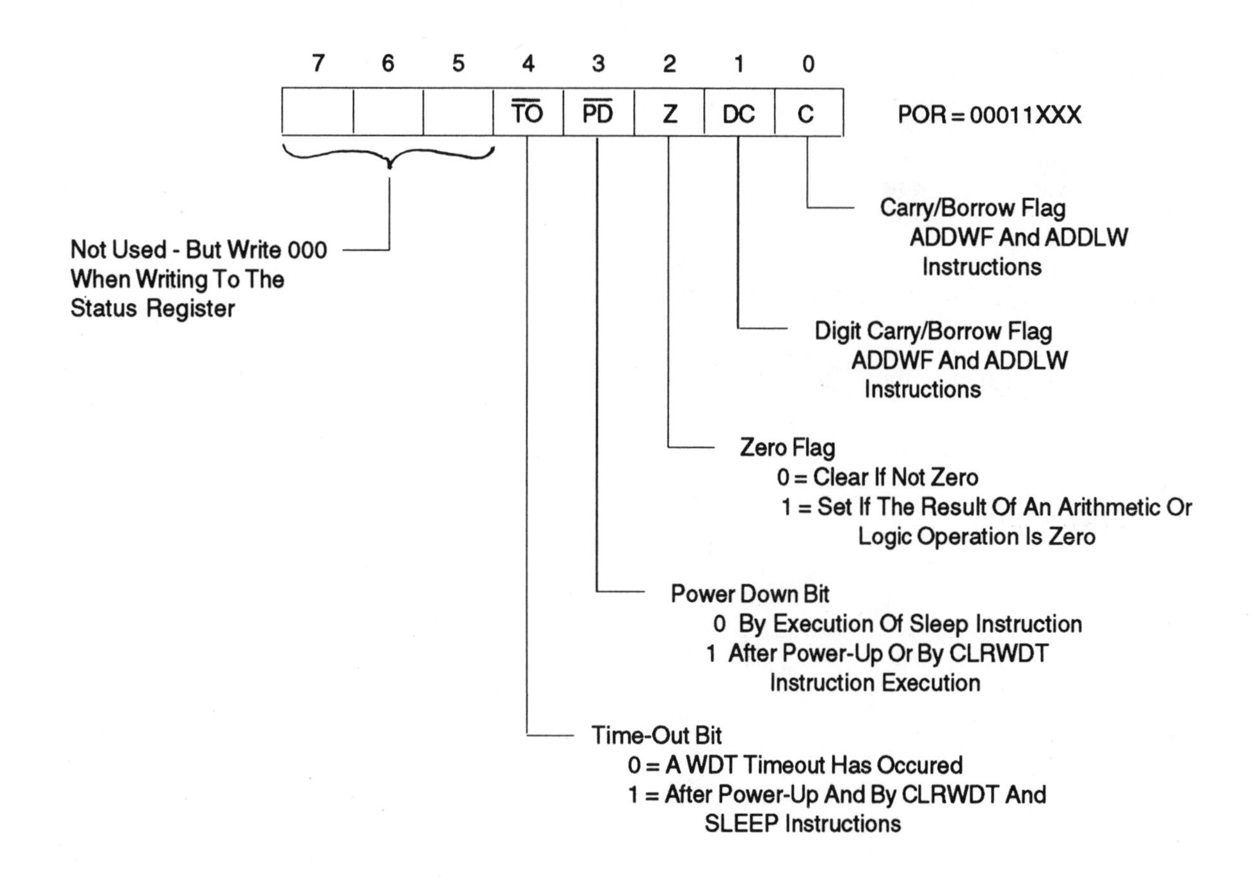

NO TIMER/COUNTER OVERFLOW INTERRUPT/FLAG OUTPUT

With the PIC16C54, the only way to find out the status of the timer/counter is to read it. The PIC16C54 can't be doing much of anything else while the timer/counter is being used for timing because it has to watch for timeout.

FEWER INSTRUCTIONS

There are four instructions which are not available:

RETURN		Return from subroutine.
RETFIE		Return from interrupt.
ADDLW	k	Add literal to W.
SUBLW	k	Subtract W from literal. Result in W.

The TRIS and OPTION registers are not addressable.

The TRIS and OPTION instructions will work as they do with the 16C84.

PIC16C54 PROGRAMMING EXAMPLE

Following is an example for the PIC16C54. It does the same thing as the first example for the PIC16C84, so you can compare them.

- Note processor type.
- No power-up timer "fuse".
- ORG statement, GOTO at 0x1FF.

```
;=======PICT1.ASM============================4/21/96==
        list    p=16c54
        radix   hex
;-----------------------------------------------------
;       cpu equates (memory map)
portb   equ     0x06
;-----------------------------------------------------
        org     0x000
start   movlw   0x00     ;load w with 0x00
        tris    portb    ;copy w tristate, port B
;                          outputs
        movlw   0x0f     ;load w with 0x0F
        movwf   portb    ;load port B with contents
;                          of w
circle  goto    circle   ;circle
;
        org     0x1ff    ;pointed to by reset vector
        goto    start
;
        end
;-----------------------------------------------------
;at blast time, select:
;       memory unprotected
;       watchdog timer disabled (default is enabled)
;       standard crystal (using 4 MHz osc for test)
;=====================================================
```

TIMING AND COUNTING

The PIC16C54 timer/counter is referred to as RTCC (real time clock counter) or, more recently as the TIMER0 (TMR0) module. I will simply refer to it as the timer/counter.

The timer/counter's features are:

- 8-bit
- Read/write
- 8-bit software programmable prescaler
- Internal or external clock
- Edge-rising or falling (external clock)
- Increments

There is no timer/counter interrupt or flag. The timer/counter must be sampled by reading f1 (0x01) to determine its status. This means that the PIC16C54 timer/counter may be used for event counting and for generating time delays. For time delays, the microcontroller must continually check the count (at least just ahead of time out) which means it can't be doing much else while timing is going on. The PIC16C84 has a timer/counter with interrupt on overflow from 0xFF to 0x00 and is capable of doing other tasks while timing/counting is going on.

An option register is associated with both the timer/counter and the watchdog timer.

- 6-bits
- Write only
- In hardware (i.e., not a file register)

Executing the OPTION instruction causes the contents of the W register to be transferred into the option register.

The clock source may be either the PIC16C54's internal instruction cycle clock or the RTCC/T0CKI pin. An external clock may be an oscillator running (much) slower than the PIC16C54's clock oscillator or it might be a source of pulses to be counted.

External clock pulses may be detected on their rising or falling edge (software selectable).

The prescaler is assigned to either the timer/counter or the watchdog timer under software control by using the OPTION instruction.

The prescaler value may be 1-of-8 as determined by the bits in the option register bits 2,1,0 (see table). Timing/counting may be done without the prescaler by assigning the prescaler to the watchdog timer (the only way to do it).

When the prescaler is assigned to the timer/counter, all instructions which write to the timer/counter register f1 (0x01) will clear the prescaler.

```
CLRF
MOVWF
BSF
Etc.
```

If an external clock source is used with no prescaler, synchronization of the external clock input must take place. This requires a short sampling procedure plus a delay after synchronization occurs and prior to the timer/counter being incremented.

There are some requirements for an external clock signal.

No Prescaler

Input high for at least 2 Tosc
Input low for at least 2 Tosc

With Prescaler

Input period of at least 4 Tosc divided by the prescaler value
Highs and lows must be of greater than 10 nanoseconds duration

Tosc is the period of the PIC16C54 clock oscillator.

If there is a write to the timer/counter, incrementing is inhibited for the next 2 instruction cycles. This can be compensated for by adjusting the number loaded in the timer/counter.

TIMER/COUNTER DESCRIPTION

The timer/counter is an 8-bit software programmable counter. It's input is the PIC16C54's internal clock or an external source via the timer/counter pin fed through an 8-bit software programmable prescaler to the counter. The only output is reading the timer/counter register. (0x01).

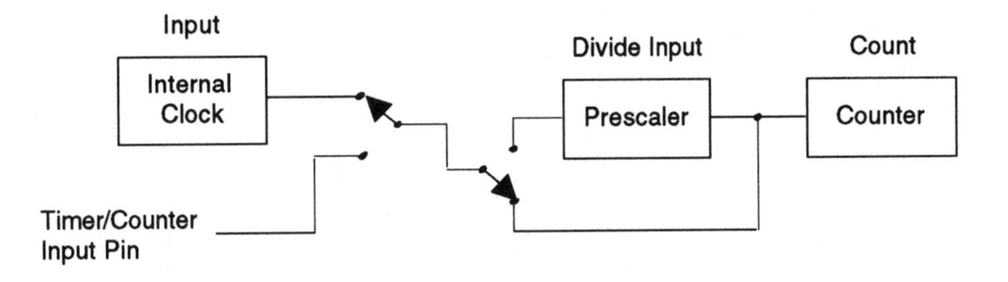

The prescaler is used to divide the input by:

```
  1  (bypass scaler by assigning it to the watchdog timer)
  2
  4
  8
 16
 32
 64
128
256
```

The timer is incremented by incoming pulses. When the count climbs through 0xFF, the count starts over at 0x00. The timer may be incremented over and over if need be and the number of times the counter reaches a certain value may be counted.

USING THE TIMER/COUNTER

Setting up the timer/counter

- Assign prescaler.
- Set up timer/counter by sending the proper bits to the option register.

Starting the timer/counter

Clear the timer/counter file register.

Counter

Counter counts up and will overflow when the count reaches 0xFF. After that, the counter contents mean nothing because there is no way to know an overflow has occurred.

How do we know the timer is doing something?

Successive reads of the timer/counter file register or test a bit in the timer/counter file register.

Timer/counter will keep counting as long as:

- It is not cleared or written to by program instructions.
- The microcontroller is not reset.

Stopping the timer

Can't - it just runs and overflows each time the count reaches 0xFF.

Prescaler

Divides the clock input by one-of-eight values which effectively reduces the frequency of the clock.

TIMER/COUNTER EXPERIMENT

This one is your deal! Create an on/off output using the timer/counter. Use a read loop to watch the timer/ counter until it increments to the value you have chosen.

MENDING OUR WAYS

SINK VS. SOURCE

The circuit examples which include LED's show them connected in series with a current limiting resistor to ground. A logic "1" turns on the LED. This method was chosen because it is intuitive for the beginner to have a "1 " turn on the LED. Current is sourced by the port line with this arrangement. A "proper" design would is shown below:

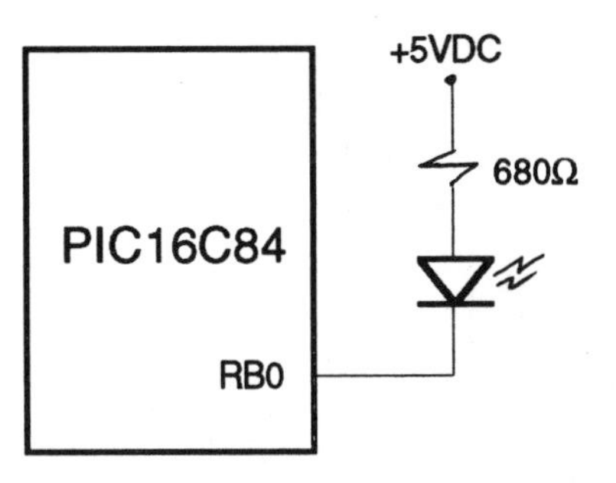

The port line sinks current and logic "0" turns on the LED. CMOS devices are better at sinking current than sourcing it. The maximum sink and source current by any I/O pin is 25 mA and 20 mA respectively. The maximum sink or source current for a port at any given time is as follows:

	Port A	Port B
Sink current	80 mA	150 mA
Source current	50 mA	100 mA

FILE REGISTER BANK SWITCHING

The option and TRIS registers are not really buried after all. There are two banks of file registers, bank 0 (the one you know about) and bank 1.

File Registers

	BANK 0	BANK1	
0x00	Indirect Address*	Indirect Address*	0x80
01	TMR0	Option	81
02	PCL	PCL	82
03	Status	Status	83
04	File Select	File Select	84
05	Port A Data	TRISA	85
06	Port B Data	TRISB	86
07			87
08	Ignore	Ignore	88
09	Ignore	Ignore	89
0A	PCLATH	PCLATH	8A
0B	INTCON	INTCON	8B
0C			8C
	36 General Purpose File Registers	Mapped In Bank 0 **	
0x2F			0xAF

* Not physically implemented
** Example: A write to 0x8C will result in data appearing at 0x0C

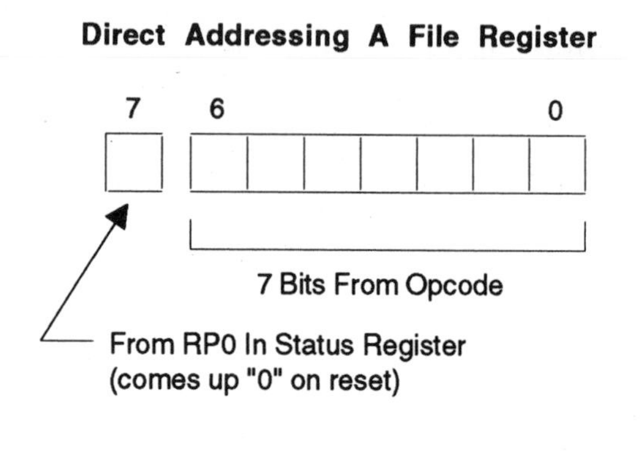

The control bit for bank switching is located in the status register.

```
RPO   bit 5
-----------------
        0 = bank 0
        1 = bank 1
```

Notice that some of the special purpose file registers appear in both banks. They will be accessed regardless of which bank is selected at the time. If RP0 is 1 and a general purpose file register is accessed, bank 0 will be accessed. The most significant bit of the direct address will be ignored.

File registers 0x08, 0x09, 0x88 and 0x89 have to do with programming the 16 x 8 EEPROM data memory which has it's own address space (beyond the scope of this book).

Soooo ..... if you want to write to the option register directly instead of using the ("illegal") OPTION instruction:

```
status      equ         0x03
rpo         equ         5
;
            bsf         status,rp0    ;switch to bank 1
            write       option        ;flip bit(s) in option register
            bcf         status,rp0    ;switch back to bank 0
```

If you want to write to the port B tristate register directly instead of using the ("illegal") TRIS instruction:

```
status      equ         0x03
rp0         equ         5
trisb       equ         0x86
;
            bsf         status,rp0    ;switch to bank 1
            movlw       b'00001111'   ;inputs vs. outputs
            movwf       trisb
            bcf         status,rp0    ;switch back to bank 0
```

BANK SWITCHING DEMO

The time delay demo program from page 108 is modified here to show how bank switching works.

```
;=======PICT24.ASM===========================8/11/96==
;time delay demo
;       tmr0, internal clock divided by 256
;       watch tmr0, bit 5 = count to 33
;       delay = 8.4 milliseconds
;----------------------------------------------------
        list    p=16c84
        radix   hex
;----------------------------------------------------
;       destination designator equates
w       equ     0
f       equ     1
;----------------------------------------------------
;       cpu equates (memory map)
tmr0    equ     0x01
status  equ     0x03
portb   equ     0x06
count   equ     0x0c
option  equ     0x81
trisb   equ     0x86
;----------------------------------------------------
```

```
;       bit equates
rp0     equ     5
;------------------------------------------------------
        org     0x000
;
start   clrwdt          ;prep for assign prescaler
        bsf     status,rp0  ;switch to bank 1
        movlw   b'11010111' ;assign prescaler,
;                   internal clock, divide by 256
        movwf   option  ;write to option register
        movlw   0x00    ;load w with 0x00
        movwf   trisb   ;write to tristate B, port B
;                   outputs
        bcf     status,rp0  ;switch to bank 0
        clrf    portb   ;all lines low
go      bsf     portb,0 ;turn on LED
        call    delay   ;delay via sub
        bcf     portb,0 ;turn off LED
        call    delay   ;delay via sub
        goto    go      ;repeat
;
delay   clrf    tmr0    ;clear TMR0, start counting
again   btfss   tmr0,5  ;bit 5 set ?
        goto    again   ;no, clear, again
        return          ;yes, end delay
;
        end
;------------------------------------------------------
;at blast time, select:
;       memory unprotected
;       watchdog timer disabled (default is enabled)
;       standard crystal (using 4 MHz osc for test)
;       power-up timer on
;======================================================
```

In the second line of the program the BSF instruction is used to set the RP0 bit (bit 5) in the status register which selects file register bank 1. While bank 1 is selected, the option register is written to (select timer options) and the TRISB register is written to (port B data direction). Then the RP0 bit is cleared to switch back to bank 0. All this is straight forward with one exception! The file register memory map for the PIC16C84 shows the follow addresses:

Register	Bank	Address
option	1	0x81
trisb	1	0x86

Using these addresses in the CPU equates results in two MPASM assembler warnings. These warnings relate to the two registers and say "Argument out of range---least significant bits used." These warnings can be ignored, but it useful to know why they appear.

When direct addressing a file register, the low-order 7 bits come from the opcode (address label in the instruction) and the 8th bit (most significant bit = bit 7) comes from the RP0 bit in the status register.

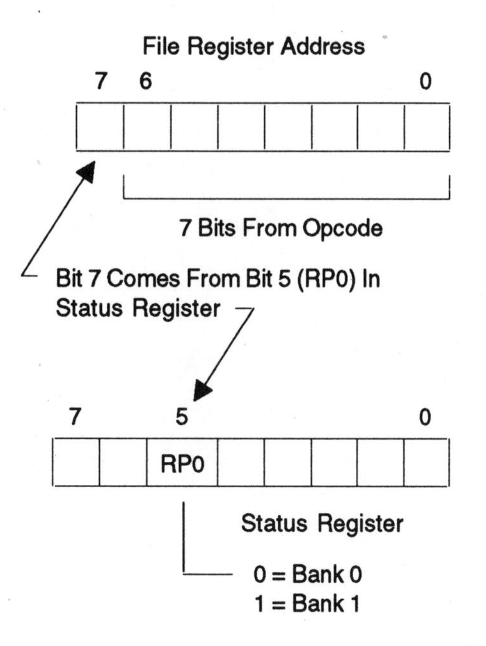

Sooo .... in the case of TRISB, the address 0x86 used with the instruction (trisb equ 0x86) will supply the low order 7 bits 0000110 and the RP0 bit will supply bit 7=1 giving the address 10000110 = 0x86.

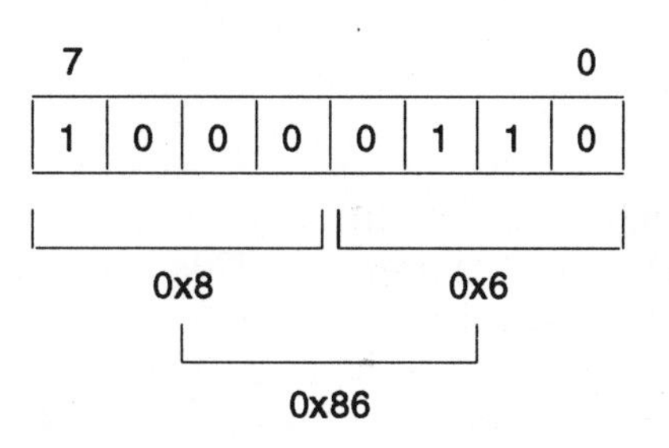

The assembler warnings may be avoided by using a technique which is beyond the scope of this book. Simply ignore them for now.

INTERRUPTS AND BANK SWITCHING

From the point of view of saving context when an interrupt occurs, the status register will be accessed whether in bank 0 or 1 at time of interrupt. Restoring the contents of the status register at the end of the interrupt service routine will take care of selecting the proper file register bank prior to resumption of main line code execution.

In terms of the interrupt service routine itself, it would be well to switch to bank 0 at the start of the routine in case the microcontroller was in bank 1 at the time the interrupt occurred. This should be done if your program includes bank switching.

```
status      equ         0x03
rp0         equ         5
;
            bcf         status,rp0   ;select bank 0
```

File register bank switching is not used in the PIC16C54 (FSR 7,6,5 are not implemented). File register bank switching is used in :

PIC16C57, 58
PIC16C7X
PIC16C8X

PROGRAM MEMORY PAGING

Program memory is divided into pages as follows:

```
                          12-bit Core      14-bit Core
                          Base-line        Mid-range
                          --------------------------
      Memory Locations       512              2K
         Per Page
```

Program memory page selection may be required (assuming your program must occupy more than page 0) by means of control bits in a register. This is beyond the scope of this book, but be aware that this requirement exists.

The PIC16C84 has 1K of program memory (less than 2K, mid-range part) and program memory paging is not required. The PIC16C4 has 512 words of program memory and program memory paging is not required. The PIC16C57, 58 (base-line parts) have more than 512 words of program memory, so paging is required for programs of more than 512 words.

INCLUDE FILES

The processor-specific equates which are used repetitively may be placed in a file called an "include" file which may be included in the assembly process. This saves creating the equates each time you write a new program. Using the time delay demo program again as an example, the destination designator, CPU and bit equates are removed from the program listing and placed in a separate include file named P16C84.INC.

Program:

```
;=======PICT25.ASM============================8/12/96==
;time delay demo
;       tmr0, internal clock divided by 256
;       watch tmr0, bit 5 = count to 33
;       delay = 8.4 milliseconds
;-----------------------------------------------------
        list    p=16c84
        radix   hex
        include <P16CXX.INC>
;-----------------------------------------------------
        org     0x000
;
start   clrwdt          ;prep for assign prescaler
        bsf     status,rp0  ;switch to bank 1
        movlw   b'11010111' ;assign prescaler,
;                       internal clock, divide by 256
        movwf   option  ;write to option register
        movlw   0x00    ;load w with 0x00
        movwf   trisb   ;write to tristate B, port B
;                          outputs
        bcf     status,rp0  ;switch to bank 0
        clrf    portb   ;all lines low
go      bsf     portb,0 ;turn on LED
        call    delay   ;delay via sub
        bcf     portb,0 ;turn off LED
        call    delay   ;delay via sub
        goto    go      ;repeat
;
delay   clrf    tmr0    ;clear TMR0, start counting
again   btfss   tmr0,5  ;bit 5 set ?
        goto    again   ;no, clear, again
        return          ;yes, end delay
;
        end
;-----------------------------------------------------
;at blast time, select:
;       memory unprotected
;       watchdog timer disabled (default is enabled)
;       standard crystal (using 4 MHz osc for test)
;       power-up timer on
;=====================================================
```

Include file:

```
;=======P16C84.INC===========================8/12/96==
;time delay demo include file
;-----------------------------------------------------
;       destination designator equates
w       equ     0
f       equ     1
;-----------------------------------------------------
;       cpu equates (memory map)
tmr0    equ         0x01
status  equ         0x03
portb   equ         0x06
count   equ         0x0c
option  equ         0x81
trisb   equ         0x86
;-----------------------------------------------------
;       bit equates
rp0     equ      5
;=====================================================
```

Notice the include statement in the program listing (PICT25.ASM). Assembling PICT25.ASM will include the file P16C84.INC.

Include statements may be in one of the following forms:

```
include "filename.ext"
```

or

```
include <filename.ext>
```

The include file must be in the current working directory (such as \PIC) at assembly time.

You may want to consider creating a complete include file of your own for the PIC16C84 to include:

- Destination designator equates
- CPU equates (memory map)
- File registers
- Bit equates
    - Option register
    - Status register
    - INTCON register

As an alternative, an include file called P16CXX.INC comes with MPASM. I used it successfully with PICT25.ASM. You will want to print out the list of labels used in P16CXX.INC and use them if you go this route.

I/O CONVERSION

INPUT CONVERSION

Input port lines must be able to clearly distinguish a "1" from a "0". If a signal is making a transition from LO to HI, the input port must see one or the other. The snap-action feature of the schmitt trigger will solve this problem. The transition from LO to HI will be clean regardless of how slowly the input is changing. Once switching has occurred, a large change in input voltage is required to switch back.

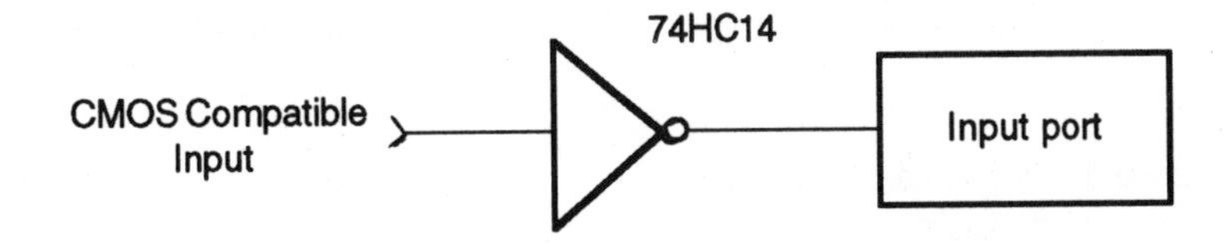

The 74HC14 is a hex schmitt trigger inverter.

In a situation where the supply voltage for the input circuit is greater than 5 volts, a 2N7000 MOSFET (Motorola or Siliconix) may be used. A voltage divider on the input presents the proper level to the gate of the MOSFET.

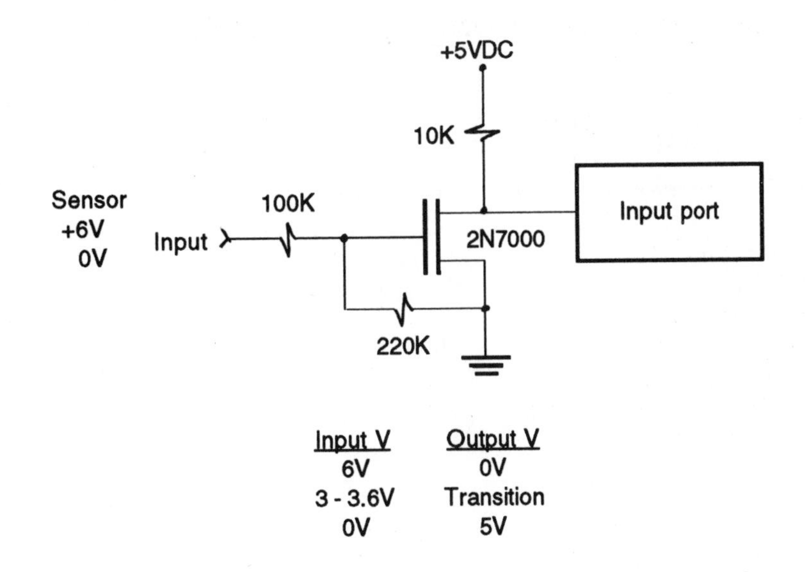

Input V	Output V
6V	0V
3 - 3.6V	Transition
0V	5V

If the input signal is not 5 volt CMOS compatible, some voltage level conversion and perhaps some isolation may be required. An optocoupler will serve the purpose. It consists of an LED on the input and a phototransistor on the output.

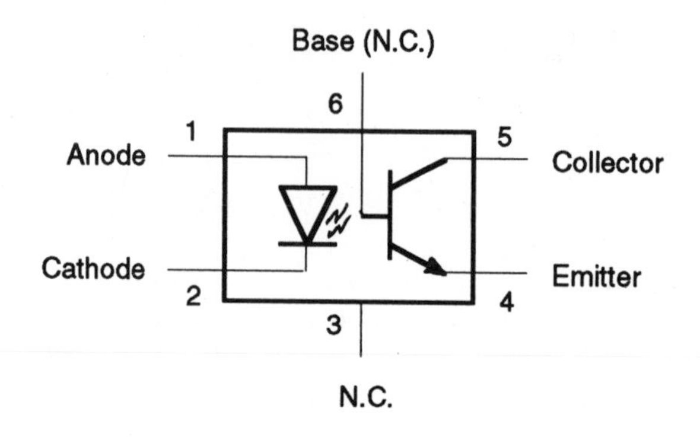

Since the coupling of output to input is via a light beam, electrical isolation of the output and input is achieved. Optocouplers are rated in KV (isolation voltage). The 4N26 (Texas Instruments) is readily available and offers the following specifications:

```
High voltage isolation                  1.5 KV
Max collector-emitter voltage            30 V
Max emitter-base voltage                  7 V
Max input diode forward current          80 mA
Max collector current                    -- mA
```

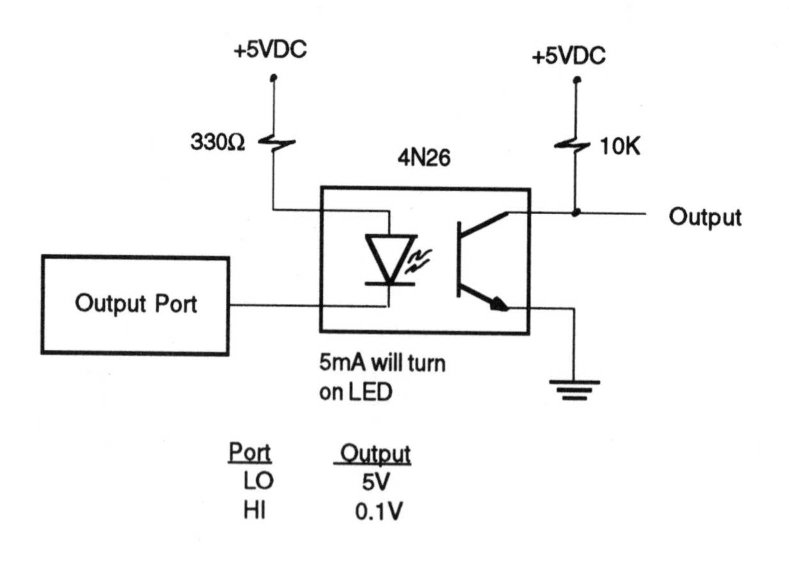

A simple application example might be sensing a switch closure in a 24 VDC system.

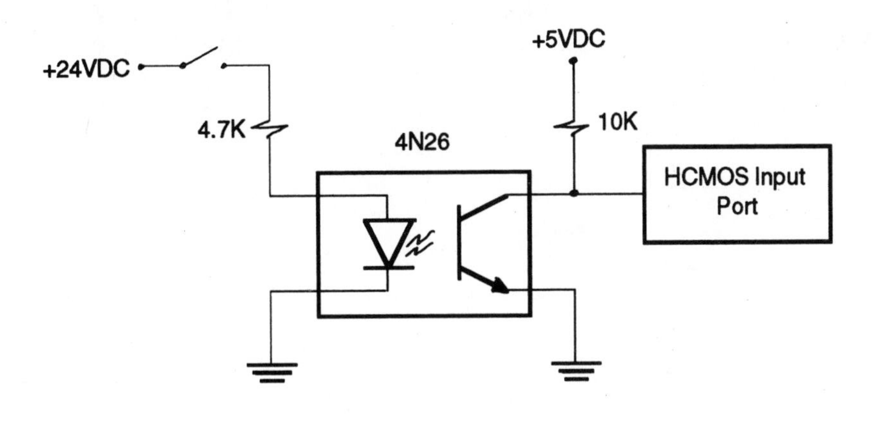

OUTPUT CONVERSION

It is commonly necessary to drive loads larger than HCMOS output ports can handle. Various devices can be used to boost the output of a port chip such as a transistor, MOSFET, inverter/buffer, optocoupler or solid state relay. Another possibility is to use a port chip with greater drive capability built-in.

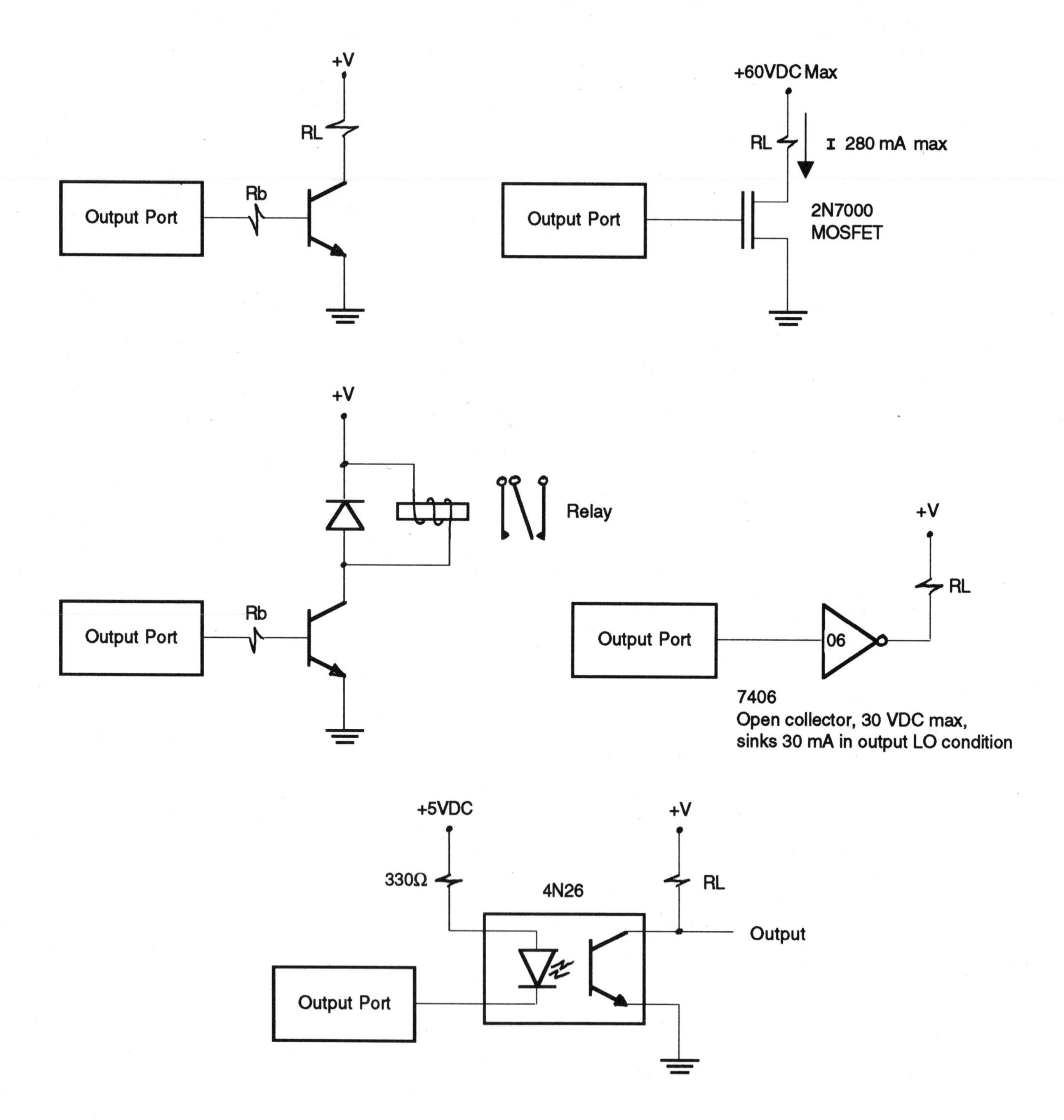

The Sprague ULN-2803A Darlington array provides eight channel high voltage, high current capability. The inputs are 5V TTL/CMOS compatible while the outputs can handle 500 mA (max) at up to 50 volts (max). Each channel has an open-collector output (Darlington pair) with an integral diode for inductive load transient suppression.

ULN-2803A Darlington Driver

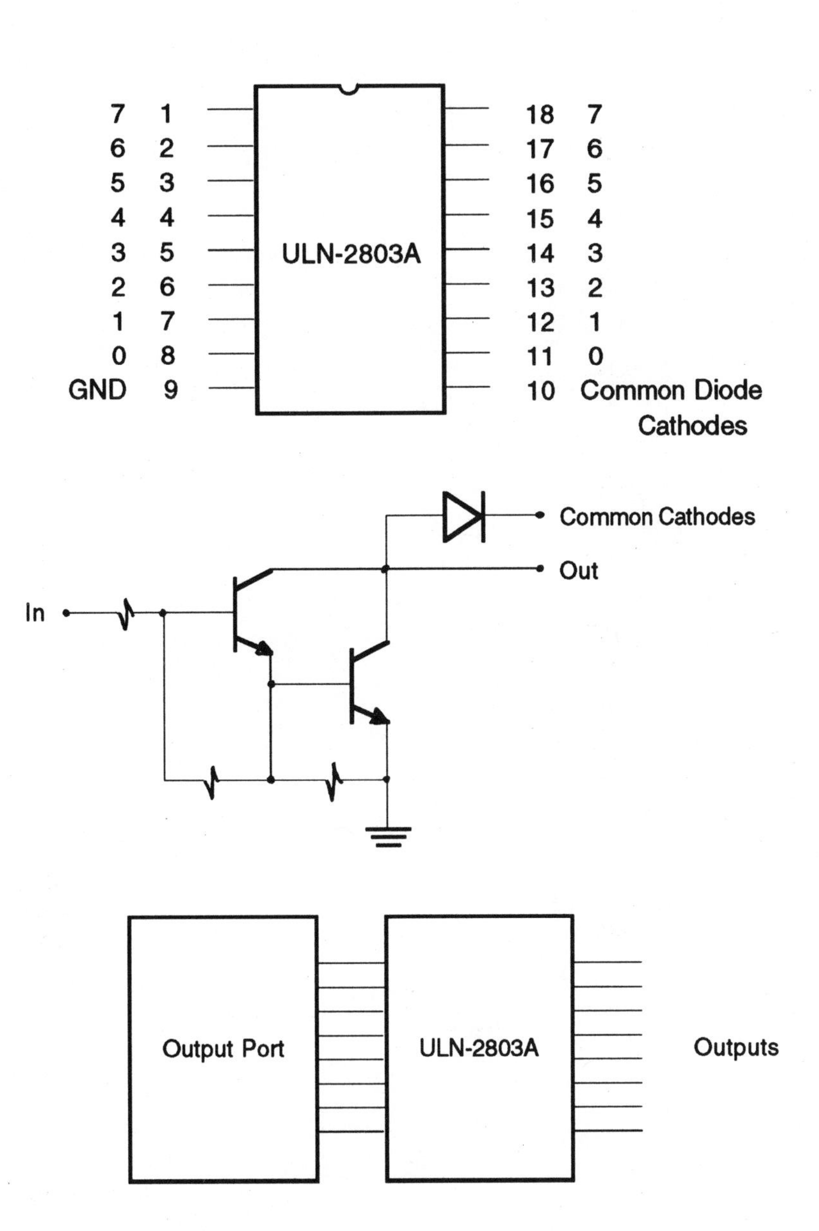

A variety of optocouplers are available for output conversion applications. The light activated output component may be a transistor, Darlington transistor pair, field effect transistor, Schmitt trigger, SCR or TRIAC.

Solid state relays use optocouplers to provide control-to-load isolation. The output is connected to a transistor for DC applications or to an SCR or TRIAC for AC applications. A variety of solid state relays are available. Choose an HCMOS compatible input with output to suit the application.

APPENDIX A
SOURCES

Following is a list of sources for PIC16/17 parts, information and tools.

```
Microchip Technology Inc.                    Manufacturer of PIC16/17
2355 West Chandler Blvd.
Chandler, AX  85224-6199   USA

(602) 786-7200

http://www.microchip.com


microEngineering Labs                        Programmers and software
Box 7532
Colorado Springs, CO  80933   USA

(719)520-5323


ITU Technologies                             Programmers and Software
3477 Westport Ct.
Cincinnati, OH  4528-3026

(513)574-7523
http://www.itutech.com


DonTronics                                   Programmer and other
P.O. Box 595                                    PIC16/17 related
Tullamarine 3043   Australia

Int+ 613 9338-6286
Int+ 613 9338-2935 FAX
http://www.labyrinth.net.au/~donmck


Digi-Key Corporation                         Parts
701 Brooks Ave. South
Thief River Falls, MN 56701-0677   USA

(800)-344-4539
```

APPENDIX B
HEXADECIMAL NUMBERS

Binary numbers which are two bytes long are difficult to recognize, remember and write without errors, so the hexadecimal numbering system is used instead. Think of hex as a kind of shorthand notation to make life easier rather than some kind of terrible math.

Hexadecimal	Binary	Decimal
0	0000	0
1	0001	1
2	0010	2
3	0011	3
4	0100	4
5	0101	5
6	0110	6
7	0111	7
8	1000	8
9	1001	9
A	1010	10
B	1011	11
C	1100	12
D	1101	13
E	1110	14
F	1111	15

Hexadecimal is the language used in this book. It will seem awkward at first, but working with binary is time consuming and will result in errors which would quickly force you to learn hexadecimal. Using hexadecimal is not difficult. All you need is a little practice.

One byte requires two hex digits. Note that the bits representing a byte are sometimes shown in groups of four. Note, also, that the most significant binary digit is on the left.

Hex numbers are denoted by "0x" in this book.

P.O. Box 501
Kelseyville, CA 95451

Voice (707) 279-8881 FAX (707) 279-8883 EMAIL squareone@zapcom.net

ORDER FORM
Easy PIC'n
MICROCONTROLLER BOOK

Date ____________

Name ______________________________ Phone (____) ________________

Shipping Address

____
____
____

We accept VISA, MC, AE, Check, MO

Credit Card Information:

Type ____
Number ____
Expiration Date ____
Name On Card ____

Credit card holder address if different from shipping address

____
____

	Qty.	$ Ea.	$ Total
Easy PIC'n Microcontroller Book	___	29.95	________

Merchandise Total	________
CA Sales Tax (7.25 %) CA Residents	________
Shipping/Handling Charges	________
Total	$ ________

$3.00 U.S. Priority Mail - Outside USA $6 to $14 U.S. depending on country